# Health, Safety, and *and* Environmental Data Analysis

*A Business Approach*

## Anthony J. Joseph

**CRC Press**
Taylor & Francis Group
Boca Raton  London  New York

CRC Press is an imprint of the
Taylor & Francis Group, an **informa** business

| | |
|---|---|
| Acquiring Editor: | Ken McCombs |
| Editorial Assistant: | Susan Alfieri |
| Project Editor: | Helen Linna |
| Marketing Manager: | Arline Massey |
| Direct Marketing Manager: | Becky McEldowney |
| Cover design: | Dawn Boyd |
| PrePress: | Kevin Luong |

CRC Press
Taylor & Francis Group
6000 Broken Sound Parkway NW, Suite 300
Boca Raton, FL 33487-2742

First issued in paperback 2019

ISBN-13: 978-1-56670-233-1 (hbk)
ISBN-13: 978-0-367-40080-4 (pbk)

Library of Congress Card Number 97-73212

**Library of Congress Cataloging-in-Publication Data**

Anthony Joseph
      p.   cm.
     Includes bibliographical references and index.
     ISBN 1-56670-233-X1.
     1. Environmental science—statistical methods.    2. Industrial hygiene—statistical methods.
     3. Occupational health—statistical methods   I. Joseph, Anthony. II. Title
     TD426.M62  1997
     628.5'5—dc20

                                        97-73212
                                         CIP

**Visit the Taylor & Francis Web site at**
**http://www.taylorandfrancis.com**

**and the CRC Press Web site at**
**http://www.crcpress.com**

# Contents

## Section A   Basic Statistical Concepts

## Section B   Data Gathering and Analysis

## Section C    Information Generation

# *Preface*

The information presented in this book is based on a series of lectures delivered by the author to students at college and groups of adults pursuing continuing education classes in Public Health, Occupational Safety and Health, and Environmental Studies. In today's world, statistics is an integral subject in any scientific study.

This book is written especially for persons who use statistics on their jobs and are not statistically inclined. To them, I strongly recommend that an initial surge of interest and enthusiasm must be generated before studying the subject. This book will also be useful for college students pursuing programs in safety, industrial hygiene, public health, and environmental protection. The experiences I gained in delivering these lectures are incorporated into this book. The information presented is based on several texts listed in the selected bibliography, reports, journal articles, and experience of the author. The subjects are approached in a simple and factual manner, complemented with examples. Mathematical derivations are omitted, except where the comprehension of the idea is dependent on the derivation. Although great efforts were made to ensure simplicity, clarity, and continuity, this book, like other statistics texts, has in some ways been fragmented. The text enunciated the principles and commonly employed methods of statistics by proceeding from one topic to another in a hopefully logical sequence, with examples from published studies, case studies, or text books. These examples serve as illustrations.

This book is intended for practicing EHS professionals, students pursuing a career in occupational health and safety and environmental studies, and managers of safety, health or environment. It is divided into three sections:

1. Basic statistical concepts
2. Data gathering and analysis
3. Information generation

Knowledge of basic statistical concepts such as presentation of data, measurements of location and dispersion, and elementary probability and distributions are essential to applying statistical techniques and methods. These are covered in Chapters 1 to 3.

Data gathering and analysis topics such as, sampling methods, sampling theory, testing and interference are essential for knowing the principles and basic methods of statistics. These will also provide skills to evaluate published numerical information critically, evident in most EHS reports. These are covered in Chapters 4 to 6.

Information generation topics such as, regression and correlation analysis, time series, linear programming, network and Gnatt charting, and decision analysis, are presented as tools that can be used to convert data into meaningful information. These are covered in Chapters 7 to 11.

Chapter 12 is a special chapter, it features six examples of projects that were successful because a statistical approach was adopted. Whatever statistical understanding you may have gained from reading this book or consulting the listed texts in the selected bibliography, synthesis of experience, understanding of statistics, and knowledge of your profession must occur.

All efforts were made to ensure that this book covers the statistical ideas necessary for the practicing environmental, health, and safety professional who will like to use statistical data to communicate information with a business approach effectively. I sincerely hope that by your reading this book, the statistical aspects of your job will be more meaningful and rewarding.

Have fun!
Anthony J. Joseph

# *About the Author*

**Anthony J. Joseph** has twenty years of experience in the field of occupational health and safety. He is currently a professor at the Indiana University of Pennsylvania, one of the premier health and safety schools in the country. Professor Joseph also holds a Ph.D. in Environmental Engineering, an M.S. in Environmental Engineering, Pollution Control, as well as in Safety Sciences.

# An Overview on
# The Importance of Statistics to
# Environmental, Health and
# Safety Professions

Environmental, Health, and Safety (EHS) professionals are engaged in the collection of data on their jobs, such as data related to lost-time accidents and emission levels of sulfur oxides. The tendency is to collect and store the data to satisfy legal requirements. The wealth of information embodied in these collections of data is underutilized or seldom used. An appreciation and understanding of statistical methods are essential for data collection to be meaningful. We live in a world where statistics influence our lifestyles and behavior. Our life is affected by human events such as illness, accidents, and environmental disasters. Most of these events are recorded and documented in numerical format. These collections of data and information over a period of time are the body of knowledge often called statistics. As with other words, the word "statistics" has different meanings to different people, for example it means mathematics to the lay person, and the study of data to the statistician.

A narrow view of the subject matter of statistics is that it simply involves page upon page of numbers in volumes upon volumes stored on the computer or on the shelf. This is only partly true. In the broader sense, statistics include techniques for tabulating and graphing data for presentation, and methods of summarizing and analyzing data. Statistics can be divided into two groups, called descriptive and inferential. Vital statistics such as birth, death, marriage, divorce, and the occurrence of communicable diseases, are used frequently in the EHS profession. Inferential statistics is the logical basis by which conclusions regarding populations are drawn from results obtained in a sample. This process of inference from sample to population pervades the fields of safety, health, environmental science, medicine and social studies. Consider the following example of statistical inference:

> A worker shows signs of lead poisoning, as a result, a sample of blood or urine or a biopsy of tissues is taken. From the sample obtained from the patient, a conclusion is drawn regarding a larger "population;" namely, the patient's total urine or blood volume, or his entire organ.

This notion of inference from sample to population has as its underlying foundation, the mathematical theory of probability. This does not mean that one must know the mathematical theory to use the statistical methods effectively. What it means is that one must understand the basis of the methodology, the assumptions governing the use of the techniques and, the proper interpretation of the results. If I can draw the analogy to the operating of a car, one need not know the mechanics of internal combustion to be able to drive a car safely. The attainment of particular operational skills along with knowledge of the rules of the road are sufficient for operating the car safely, and effectively in most countries.

The application of statistical results is so widespread that the importance of statistics can hardly be overemphasized. Rates, ratios, and probabilities are all related to statistics, for example infant mortality rates, that is the number of deaths less than one year of age reported during the same year.

A branch of medicine called epidemiology is based on statistics. This branch of medicine contributed and is still contributing to the improvement of public health. The developments of immunology and the acquisition of new knowledge regarding the transmission of diseases were ushered by epidemiology. For example in 1854, John Snow proved that cholera was transmitted in water and not by personal contact from statistical analysis, while the identification of the cholera organism was not made until 27 years later.

Can statistics provide all the answers to our safety, health and environmental problems? Statistics deal with measurable aspects of things. Therefore, it can seldom give the complete solution to a problem. Statistics can provide a basis for judgement and decision making. The limitations of the data collected and use must be fully delineated. Clearly, all observations must be accurately recorded, collated, analyzed and presented. Usually, it is this body of data transformed into information that is used to predict future events, to infer or to relate, or to classify events that impact on our lives.

The age of computers has brought us the ability to amass volumes of data and information. However, the extensive use of this data gathering marvel to transform data into useful information has not in most cases cured the malady that exists among many EHS professionals. Today we are enjoying better health and safety in our work places, home and communities because of the "body of knowledge" obtained from data and observations. The development of vaccines resulted from statistical studies. Just reflect for a moment on the important of statistics in EHS. It is impossible to be a professional in EHS without knowing basic statistics.

# List of Figures

# List of Tables

*Section A*

# Basic Statistical Concepts

# 1

## Presentation of Data

### 1.1  Introduction

Statistics involves the collection and presentation of data in different forms. Data are the numbers and observations collected, analyzed, and summarized. Elements are the entities on which data are collected. A variable is a characteristic of interest for the elements. The type of statistical analysis and presentation appropriate for the data on a particular variable depends upon the scale of measurement used for the variable. There are four scales of measurement: nominal, ordinal, interval, and ratio. Nominal scales are simply labels used to identify an attribute of the element; for example, sex — male or female. Ordinal scales are ordered or ranked data, such as used in rating the risk of a hazard — imminent, most likely, and unlikely. Interval scales are numeric; for example, lost-time days. Ratio scales are the ratio of two interval values with a defined zero; for example, workers compensation rates. Interval and ratio scales permit more rigorous statistical analysis than nominal and ordinal. Data can also be classified as qualitative or quantitative: qualitative data provide labels, and quantitative say how many or how much.

### 1.2  Presentation

The ability to select the appropriate display form is just as important as collecting the data. Undertaking any statistical presentation of data or enquiry requires consideration of certain initial matters as follows:

1. A precise definition of objectives
2. Collection of data in the right units
3. The accuracy of data required
4. A clear definition of the population
5. Cost effectiveness

These will ensure that all necessary data are collected, and consequently, time and money will not be wasted on superfluous materials.

Descriptive statistics is primarily concerned with methods of organizing, summarizing, and presenting data. Environmental Health and Safety (EHS) data collections are often very voluminous. The collection task must be viewed as a necessary first step in the decision-making or planning process of the company for growth and profitability. For example, the safety manager recommended changes to the operation because of high accident rates and associated costs by collecting 15,000 items of data in three years. Think about past occasions on your job when you failed to convince management because of lack of supporting information or data. Whatever example you can now think of, you must clearly present the data in an effective manner supportive of a major point. Briefly, data are first collected, then arranged or grouped to reduce volume, and finally expressed or presented in terms or measures that describe some behavior defined in the objectives of the exercise.

## 1.3   Accuracy and Errors

In presenting numeric data, the need for approximation comes about because obtaining accurate measure of items is often not possible. This will result in errors that can affect the validity and accuracy of the information presented. Rounding error occurs when an absolute measure is impossible, so we round up or down or to the nearest round amount. Table 1.1 shows the effects of rounding the hours of labor worked by ten departments before a lost-time accident occurred.

Assuming a bonus of $1 per employee for every safe man-hour worked was offered, consider the impact on your budget allocation of funds if you were to use rounded totals for planning. The maximum rounding up or to the nearest 000 error in this example is 10 * 500 = 5000, while rounding down is 10 * 1000 = 10,000. The average error is normally used. It is half the maximum error e.g. 87,000 ± 2500. Rounding errors are cumulative. Showing the error in the results may be very important in presenting numeric data.

## 1.4   Arrays

An array is a sequential ordered arrangement of numeric, alphabetic or alphanumeric notations. For example, the total lost time at 10 computer work stations in January are (in minutes):

Arranging the data in an array can result in the following:

**TABLE 1.1**

Hours of Labor per Department Before a Lost-Time Accident

| Actual Hours of Labor | Rounded to Nearest 000 | Rounded Up to Nearest 000 | Rounded Down to Nearest 000 |
|---|---|---|---|
| 17,426 | 17 | 18 | 17 |
| 615 | 1 | 1 | 0 |
| 1333 | 1 | 2 | 1 |
| 7289 | 7 | 8 | 7 |
| 16,438 | 16 | 17 | 16 |
| 7858 | 8 | 8 | 7 |
| 11,673 | 12 | 12 | 11 |
| 24,184 | 24 | 25 | 24 |
| 783 | 1 | 1 | 0 |
| 467 | 0 | 1 | 0 |
| 88,066 | 87 | 93 | 83 |

$6\frac{1}{2}$   4   $9\frac{1}{2}$   5   4   8   $5\frac{1}{2}$   9   7

$9\frac{1}{2}$   9   8   8   7   $6\frac{1}{2}$   $5\frac{1}{2}$   5   4

## 1.5   Tally Marks

Tally marks is a simple form of representing data. For example, displaying the number of near misses experienced by four employees during a week:

| Employee | Near Misses |
|---|---|
| A | //// |
| B | ///// ///// //// |
| C | ///// // |
| D | ///. |

## 1.6   Frequency Distribution

### Ungrouped Frequency Distribution

If many measurements of a particular variable are taken, recording the number can construct a frequency distribution of times a value occurs. For example, the total time (minutes) spent by twenty employees in setting up their molding machine was:

$$65 \quad 69 \quad 70 \quad 71 \quad 70 \quad 68 \quad 69 \quad 67 \quad 70 \quad 68$$
$$72 \quad 71 \quad 69 \quad 74 \quad 70 \quad 73 \quad 71 \quad 67 \quad 69 \quad 70$$

If the number of occurrences is placed against each output quantity, a simple frequency distribution is produced:

Time Spent by Employees

| Time (Minutes) | Number of Employees (Frequency) |
| --- | --- |
| 65 | 1 |
| 66 | 0 |
| 67 | 2 |
| 68 | 2 |
| 69 | 4 |
| 70 | 5 |
| 71 | 3 |
| 72 | 1 |
| 73 | 1 |
| 74 | 1 |
|  | 20 |

The number of employees corresponding to a particular time is called a frequency. When data are rearranged in this way, 69 and 70 minutes are obviously the most common times an employee spent in setting up a molding machine. There are many applications for which simple frequency distributions are most appropriate; for example, showing exposure levels to a particular chemical by workers.

## Grouped Frequency Distributions

Grouping frequencies together into bands or class intervals is often convenient. The bands can be either discrete or continuous. The corresponding distributions are also either continuous or discrete.

## Discrete Distribution

The number of accidents recorded by 20 companies during one year was as follows:

| | | | |
| --- | --- | --- | --- |
| 1087 | 850 | 1084 | 792 |
| 924 | 1226 | 1012 | 1205 |
| 1265 | 1028 | 1230 | 1182 |
| 1086 | 1130 | 989 | 1155 |
| 1134 | 1166 | 1129 | 1160 |

An ungrouped frequency or a simple frequency distribution would not be a helpful way of presenting the data, because each company experienced a different number of accidents in the year. The number of accidents ranged from 792 to 1265. Dividing this range into class intervals of say 100 units, starting at 700 and ending at 1299, the number of companies recording accidents within each class interval could then be grouped into a single frequency, as follows:

| Class Interval's (No. of Accidents ) | Frequency (No. of Companies) |
|---|---|
| 700–799 | 1 |
| 800–899 | 1 |
| 900–999 | 2 |
| 1000–1099 | 5 |
| 1100–1199 | 7 |
| 1200–1299 | 4 |
| | 20 |

## Continuous Distributions
The heights of 50 different individuals were grouped as follows:

| Height (Centimeters) | Number of Individuals (Frequency) |
|---|---|
| Up to and including 154 | 1 |
| More than 154, up to and including 163 | 3 |
| More than 163, up to and including 172 | 8 |
| More than 172, up to and including 181 | 16 |
| More than 181, up to and including 190 | 18 |
| More than 190 | 4 |
| | 50 |

Some observations are best presented as continuous data; for example, heights of individuals. Care must be taken in establishing intervals. In the above example, your attention is drawn to three points:

1. A height can only occur in one interval class
2. There are two open-ended class intervals, the first and last
3. The class intervals are equal except the two open-ended intervals

Open-ended class intervals are not suitable for rigorous statistical analysis, therefore they are not recommended. These are further discussed in section 1.7.

## 1.7 Cumulative Frequency Table

A cumulative frequency table is used to show the total number of times that a value above or below a certain amount occurs. For example, suppose that the volume of waste generated in one day by each of 20 employees is as follows, in lbs.:

|    |    |    |    |
|----|----|----|----|
| 18 | 29 | 22 | 17 |
| 30 | 12 | 27 | 24 |
| 26 | 32 | 24 | 29 |
| 28 | 46 | 31 | 27 |
| 19 | 18 | 32 | 25 |

We could present a grouped frequency distribution as follows:

| Waste (lbs.) | Number of Employees (Frequency) |
|---|---|
| Less than 15 | 1 |
| 15 or more, less than 20 | 4 |
| 20 or more, less than 25 | 3 |
| 25 or more, less than 30 | 7 |
| 30 or more, less than 35 | 4 |
| 35 to 50 | 1 |
|  | 20 |

A cumulative frequency distribution table for the same data would be:

|  | Cumulative Frequency (More Than) | OR | Cumulative Frequency (Less Than) |
|---|---|---|---|
| Less than 15 | 20 |  | 1 |
| > 15 to < 20 | 19 |  | 5 |
| > 20 to < 25 | 15 |  | 8 |
| > 25 to < 30 | 12 |  | 15 |
| > 30 to < 35 | 5 |  | 19 |
| > 35 to 50 | 1 |  | 20 |

The symbol > means "greater than" and < means "less than" for example $a > b$ reads $a$ is greater than $b$.

There are several simple guidelines for constructing a frequency distribution. Note that these are merely guidelines and not hard-and-fast rules.

1. The number of classes should generally be about 10 to 20. Obviously, the use of too many classes differs little from tabulation of the raw data; too few classes may obscure essential information. In practice, beginning with several fine classes is better. These fine classes can easily be combined into broader classes. If one starts with broad classes and then wishes a finer tabulation, one must return to the raw data and re-tabulate the entire series.

2. The limits for each class must agree with the accuracy of the raw data. For example, the serum uric acid levels was recorded correct to the nearest tenth of a milligram per 100 ml. The class intervals retain this accuracy; for example, the first interval is 3.0 to 3.4 mg per 100 ml. Employing an interval of 3.00 to 3.49 mg per 100 ml would be inappropriate since the data were not recorded with this degree of accuracy.

3. Intervals of equal width are convenient and ease further computation. Equal width intervals, however, are not essential.

4. The class intervals must be mutually exclusive. For example, in a frequency distribution of age, intervals of 5 to 10 year and 10 to 15 years are not mutually exclusive. With an individual age 10, one would not know whether he falls in the 5 to 10 or the 10 to 15 year class.

5. Open-ended intervals should be avoided. If the last interval for phenol levels were quoted as "greater than 8.5 mg per 100 ml," one would not know where the observations fell — whether the observations were near 8 or perhaps as high as 10 mg per 100 ml. Although this guideline is commonly neglected, its neglect makes for a certain degree of arbitrariness and difficulty when further manipulations are done with the frequency distributions. Similarly, graphical depictions of open-ended frequency distributions have trouble.

6. For further computations from frequency distribution data, determination of the midpoint of each class is essential.

## 1.8   Tabulation of Attributes

When the information to be collected is attributes, such as gender composition of a company, the data can be best presented in a table, as illustrated in Table 1.2.

**TABLE 1.2**

Analysis of Gender — Richmond Limited, 1996

| Administrative Staff | Number of Men | Number of Women | Total |
|---|---|---|---|
| Clerical Staff | 180 | 78 | 258 |
| Junior Managers | 25 | 22 | 47 |
| Middle Managers | 8 | 2 | 10 |
| Senior Managers | 3 | 1 | 4 |
| Directors | 2 | 0 | 2 |
| Total | 218 | 103 | 321 |

The following are guidelines for presenting data in tabular form:

1. The table should be given a clear title
2. All columns should be clearly labeled
3. A total column may be presented; this would usually be the right-hand column
4. A total figure is often advisable at the bottom of each column of figures
5. Tables should not be packed with too much data so that the information presented is difficult to read

## 1.9 Graphical Displays

Instead of presenting information in a table, giving a visual display as a chart or graph might be preferable. There are several forms of charts and graphs as listed below. A brief statement on the use of these displays will follow.

- Bar charts
- Line charts
- Pie charts
- Pictograms
- Venn diagrams
- Statistical maps or cartographic maps
- Gantt charts
- Histograms
- Frequency polygons

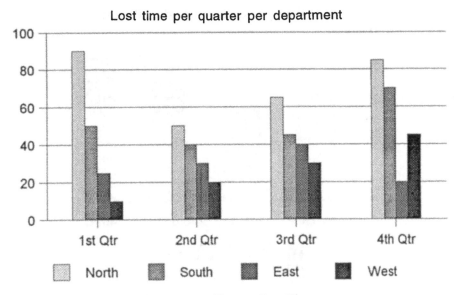

**Figure 1.1   Cluster Bar Chart.**

## Bar Charts

Bars of equal width represent the data. The height of the bar corresponds to the value of the data. There are four (4) main types: (1) cluster, (2) overlap, (3) stacked, and (4) stacked percentage.

To highlight the differences, Figures 1.1 to 1.4 are presented using the same data.

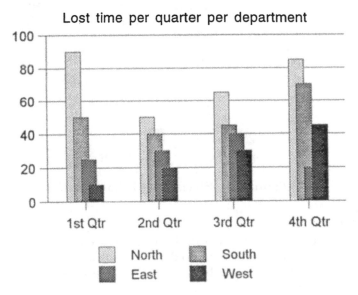

**Figure 1.2   Overlap Bar Chart.**

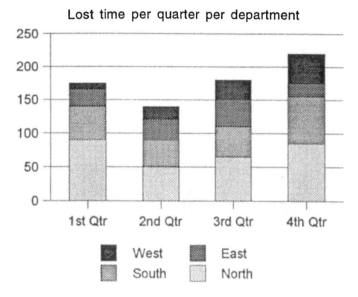

Figure 1.3    Stacked Bar Chart.

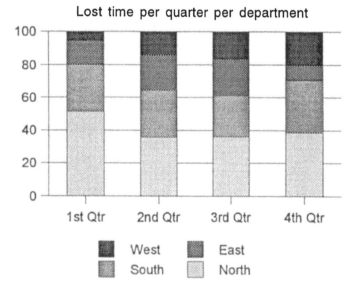

Figure 1.4    Stacked 100% Bar Chart.

## Line Charts

A line chart is similar to a bar chart. Lines are used instead of bars. The length of the line is proportional to the value represented. Line charts are often used for discrete variables. Figure 1.5 presents an example of a stacked line chart.

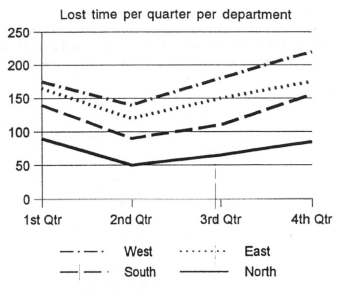

Figure 1.5   Stacked Line Chart.

## Pie Charts

Pie charts are used to show the relative size of the components of a total. This is done by drawing a circle and splitting it into sectors so that the size of a sector is proportional to the component it represents. Figure 1.6 presents an example of a pie chart.

Lost time per quarter per department

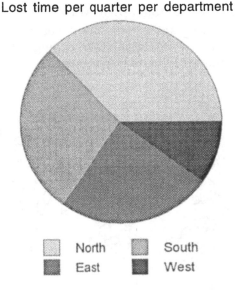

Figure 1.6   Pie Chart.

## Pictograms

The use of symbols or pictures shows values. The number of symbols should represent the values and not by symbols of different size. Pictograms are not accurate enough for serious statistical work.

## Venn Diagram

A Venn diagram is used to show subdivisions, and subdivisions of subdivisions, and so on. It is best used to show relationships. Figure 1.7 presents an example of a Venn diagram.

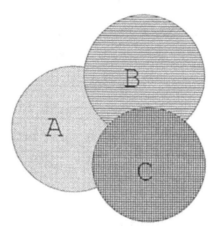

Figure 1.7    Venn Diagram.

## Statistical Maps or Cartographic Maps

Statistical maps or cartographic maps, may be used to display geographical data. A key is necessary to explain the meaning of the shadings. Figure 1.8 presents an example of a cartographic map.

Figure 1.8    Cartographic Map.

## Gnatt Charts

A Gnatt chart is a bar chart, scaled in units of time. It is sometimes used in business to show the progress, planned or already achieved, in the accomplishment of a job. A Gnatt chart can also be used to measure actual against planned achievement. The use of Gnatt charts are presented in Chapter 10 in greater detail.

## Histograms

A histogram is probably the most important type of visual display chart. It resembles a bar chart, but there are important differences: it is used when grouped data of a continuous variable are presented. The area represents the number of observations in each class covered on the chart, not by the height of the bar. Figure 1.9 presents an example of a histogram.

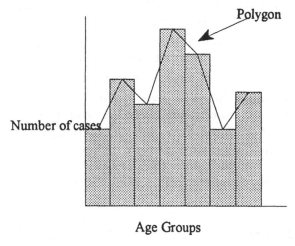

Figure 1.9  Histogram and Frequency Polygon.

## Frequency Polygons

Though less commonly used than the histogram, the frequency polygon suffices for the depiction of a frequency distribution. For each class of the distribution, one finds a point whose abscissa is the midpoint of the class whose ordinate or height is the frequency. The series of points, connected by straight lines depicts the frequency polygon, as shown in Figure 1.9.

## 1.10 Graphs

Of special interest are graphs. A graph has a horizontal axis termed the "x" axis and a vertical axis termed the "y" axis. The *x* axis is used to represent the independent variable and the $y$ axis is used to represent the dependent variable. A dependent variable is one whose value depends on the value of the independent variable; in other words, the value of $y$ depends on what the value of $x$ happens to be. In a frequency curve, $y$ would represent the number of proportion of frequencies in a class interval and $x$ would represent the class intervals. Note that time is always treated as the independent variable; when the $x$ axis on a graph represents time, the graph is usually called a "time series." Some basic guidelines follow:

1. If the data to be plotted are derived from calculations rather than measurements, include a neat table showing your calculations.
2. The scale selected should utilize as much of the graph paper as possible. Do not cramp a graph into one corner. Occasionally not starting a scale at zero is best. This is perfectly acceptable since the scale adopted is clearly shown. One way of avoiding confusion is to break the axis concerned.
3. A graph should not be crowded with too many lines; it should give a clear and neat impression of the data.
4. The axes must be clearly labeled with descriptions and units.
5. A graph must always be given a title, and where appropriate, a reference should be made to the source of data.
6. An equation such as $y = 4x + 5$ gives a straight line. This is also true of all equations of the type $y = ax + b$. When plotting straight lines, it is only necessary to calculate two points and join them up; however, three (3) points are recommended. Why?

## 1.11  Scatter Diagrams

A scatter diagram is a special type of graph useful in pointing our relationships or associations between two variables. In this type of graph, several sets of paired data are plotted on the same graph. The pattern made by the plotted points shows a possible relationship. If they tend to follow a straight line, the relationship is of a linear nature. If the pattern does not follow a straight line, the relationship may be curvilinear. If the pattern is a scatter of points, then the trend suggests that probably no relationship exists.

## 1.12  Time Value in Presentation of Data

Time value considerations in the representation of data should not be overlooked. The numerical value of data can increase or decrease, therefore influencing the financial implications of the results. Two major mathematical techniques used in finance are compounding and discounting. Compounding and discounting are antonyms. Compounding adds interest, while discounting subtracts interest. Their methods of calculation are however, similar. The basic principle of compounding is that if we invest $P now at a rate of $r$% per period, after $n$ periods it would increase to $P + s$. Discounting is the reverse: if we purchase a piece of equipment for $P, and it depreciates at a rate of $r$% per period, after n periods it would decrease to $P - s$. The basic formula is given as shown in Equation 1.1

$$S_n = P(1 + r)^n \qquad (1.1)$$

where: $P$ = original value, $r$ = rate of interest or discount, $n$ = number of periods, and $S_n$ = the value after the $n$ periods.

The good news is that there are tables called "discount tables" for different values of $r$ and $n$. It is this compounding formula used to predict future values such as those caused by inflation. The technique of discounting in finance is known as "discount cash flow" (DCF). It is used in the evaluation of capital expenditure projects to decide whether they offer a satisfactory return to the investor. The term "present value" simply means the amount or value that must be invested now for n periods at an interest rate of $r$%, to earn a future value at the end of the nth period.

Discounted cash flow involves the application of Equation 1.1 and net present value (NPV) or internal rate of return methods. The NPV is the difference between the present value of future cash inflows from a project and the present value of future cash outflows over a given period. Example: The safety director recommended installation of safety guards to the value of $18,000 to reduce the cost associated with the minor hand injuries. The projected savings based on the records per quarter are $6000, $8000, $5000, and $1000. The company requires a 10% return per quarter. How financial viable is this recommendation? Using Equation 1.1 and NPV concepts, let us examine the discounted cash flow of the recommendation as follows:

| Quarter | Cash Flow | Discount Factor | Present Value |
|---------|-----------|-----------------|---------------|
| 0 | −18,000 | 1.0 | −18,000.0 |
| 1 | 6000 | 0.9091 | 5454.6 |
| 2 | 8000 | 0.8264 | 6611.2 |
| 3 | 5000 | 0.7513 | 3756.5 |
| 4 | 1000 | 0.6830 | 683.0 |

Therefore, the NPV = −$1494.7 (the sum of column 4)

The NPV is negative which means that there is a loss. Usually projects with negative net cash flow are not approved.

The IRR method of discounted cash flow is to decide the rate of interest called the internal rate of return, at which the NPV is zero. The internal rate of return is therefore the expected earning rate of an investment. This method will show that a project is viable if the IRR exceeds the minimum acceptable rate of return.

## Depreciation and Amortization

Depreciation is an accounting technique by which the cost of a capital asset is spread over a number of different accounting periods as a charge against profit in each period. Amortization is a term sometimes used to mean depreciation; however, amortization is the process of wiping out an amount of money over a given period. There are several methods of calculating depreciation and amortization. The most common is the straight line method. Others include sum-of-the-digit (different depreciation rates each period), reducing balance (depreciation a constant percent), and sinking funds (depreciation by discounting).

As an example of the power of financial considerations incorporated into EHS proposals, a safety director recommends to management the installation of a local exhaust system as the best solution to reduce the high vapor levels in the finishing room. In so doing, the cost associated with treating workers suffering with a respiratory problem will be reduced. The estimated cost to install the system is $250,000. Management accepted his conclusions and requested him to show how the proposal will be funded. The only immediate funds available to the safety director are his annual budget allocations. The safety director presented a financial package to management in which he proposed to borrow 75% of the money at 10% annual interest rate and amortize it in two years with an IRR of 5%. This won management approval. The essential ingredients that fortified the proposal were the data collection, presentation of the information, the abilities to use the data to identifying a solution and suggest a financing plan. This is the challenge of the EHS personnel in presenting data for financial decisions.

## 1.13 Key Points in Data Presentation

1. Choose the "Presentation Tool" that is most effective in presenting the data and its purpose. Remember that some methods of presenting data call for the original data, while others require special design techniques. Within these limitations, decide upon the precise idea that you wish to communicate, then choose the method: continuous line graphs are suitable for a comparison of trends; bar charts clearly compare separate quantities of

limited number; pie charts have advantages in comparing parts with their whole; and scatter diagrams are excellent for showing tendencies.

2. Highlight one "Idea" at a time. Confine the presentation to one purpose or idea; limit the data and include only one kind of data in each presentation. Different view points on the information call for separate presentations, unless information is compared.

3. Use black-and-white for exhibits that are to be reproduced: color reproduction is expensive. Color can be adequately replaced by cross-hatching, dotted fields, continuous marks, dots, dashes, or combinations thereof.

4. Use adequate, properly located labels and titles. Labels and titles should answer the questions "what? where? and when? They should be clear, complete, and easy to understand. They should be outside the frame of the data. Only keys or legends should appear within the field of a graph or chart.

5. Give your sources. State where or how (or both), the data were obtained.

6. Use care in proposing conclusions. In particular, draw conclusions that reflect the full body of information from which the sample was taken; propose only such conclusions the data that is presented can support. Remember that tables, graphs and charts emphasize generalities at the expense of detail. Footnote in a prominent way any important detail obscured. Avoid conclusions that do not consider such distortion.

7. Consider the financial implications. The EHS concerns are important when they affect the profitability of the company. The cost benefit and the funding of the proposal are essential for obtaining management approval. Working with financial personnel on presenting data from projects is advisable.

# 2

## Measures of Location and Dispersion

### 2.1 Introduction

We can summarize and represent a set of data by a single value or values. Usually we summarize data around the central tendency in a single value. The mean, mode, median, and percentile are the simplest and most common central measures. The standard deviation and variance are the simplest measure of dispersion around the central measures or tendency.

### 2.2 Mean

There are three types of means: arithmetic, geometric, and weighted. The arithmetic mean is obtained by adding all the data values and dividing by the number of data items. The arithmetic mean of a sample of size $n$ will be denoted by $\bar{x}$ and the population of size $N$ will be denoted by $\mu$.

The geometric mean is the $n$th root of the product of the $n$ items. Mathematically expressed as:

$$\sqrt[n]{x_1 * x_2 * x_3 \cdots * x_n} \tag{2.1}$$

Weighted mean is obtained by assigning different emphasis or weights to data items, and is mathematically expressed as:

$$\frac{\sum (w_i x_i)}{\sum w_i} \tag{2.2}$$

where $w_i$ is the weight assigned to an item.

## 2.3  Median

The median is the value falling in the middle of the data set when arranged in a ranked ordered form. For grouped data, the median can be computed as:

$$L + [CI * (n/2 - a)/b] \qquad (2.3)$$

where $L$ is the lower boundary of the median class, $CI$ is the class interval, $n$ is the sample size, $a$ is the number of items in classes below the median class and $b$ is the frequency or number of items in median class.

## 2.4  Mode

The mode is the data value that occurs with the greatest frequency. In a data set it is straight forward. However, with grouped data, there are several methods. The result of the computation despite the method used is a value that will not necessarily be a value in the data set. Furthermore, computational methods will not reveal second or third modes, if they exist. Identifying the modal class in grouped data is better.

## 2.5  Percentile

The $p$th percentile is a value such that p percent of the data take on this value or less. It is a measure that locates values in the data set that are not necessarily central locations. It provides information regarding how the data items are distributed.

## 2.6  Selecting an Appropriate Measure of Central Tendency

Deciding which measure of a central tendency to use to describe your data depends largely on the distribution of the data and objectives of the presentation. For example, data that are symmetrically distributed, that is, the mean, median, and mode all occur at the same point, the choice is immaterial. A skewed distribution is one in which one tail is elongated. If it is elongated toward the right, it exhibits positive skewness, toward the left negative skewness. Also, it can be symmetrical distribution with bi-modes. Each case poses its unique challenge to the presenter. Nevertheless, the mean is the preferred measure because of its greater value in subsequent statistical computations.

## 2.7   Measures of Dispersion of Discrete Data

The costs of four accidents to one company are $5000, $3000, $100, and $100, and costs of three accidents to another company are $8000, $4000 and $150. The mean costs to both companies are an equal $4050. Information, more than just a measure of a central tendency is required if we are to investigate these performances. The dispersion of the data can provide additional information. There are three measures of interest: the range, standard deviation (square root of the variance), and the mean deviation.

### Range
The range of a data set is the difference between the highest and lowest values in the set. It depends on the extremity of the end values. When data are arrayed in ascending order, the range is simply the value of the last element minus the value of the first. The inter-quartile range is the difference between the third quartile and the first quartile. This measure of dispersion overcomes the dependency upon extreme data values.

### Standard Deviation
The standard deviation is a measure of dispersion and it is the most frequently used statistical term. The application of the standard deviation will be discussed in subsequent chapters. Mathematically, for a set of data elements, the standard deviation can be computed by:

$$\sigma = \sqrt{\frac{\sum (X_i - \mu)^2}{N}} \qquad (2.4)$$

where $\sigma$ is the standard deviation of the population and $X$, $\mu$, and $N$ are as previously defined. The sample standard deviation is obtained by substituting $X_i$ with $x_i$; $\mu$ with $\bar{x}$ and $N$ with $n - 1$. Many electronic calculators and computers are designed directly to calculate standard deviation. The square of the standard deviation is termed the variance.

Although the practical applications of the standard deviation are limited to more sophisticated statistical methods, it is interesting that there is a relationship between the standard deviation and its more simple companion measure, the range. When data are normally distributed, that is the mean, mode and median coincide, an idea described earlier, the range is approximately equal to six standard deviations. This relationship is sometimes used to make a crude estimate of the standard deviation when a more formal computation is not possible.

$$\sigma = \sqrt{\frac{\sum x_i^2 - \frac{\left(\sum x_i\right)^2}{N}}{N}} \tag{2.5}$$

Equation 2.5 can also be used to calculate the standard deviation when the mean is not known.

## Mean deviation

The mean deviation is a measure of the average amount by which the values in a distribution differ from the arithmetic mean. It is given by:

$$\text{Mean deviation} = \sum \frac{|x - \bar{x}|}{n} \tag{2.6}$$

where $|x - \bar{x}|$ is the absolute difference, such that the plus or negative differences are ignored. This measure is useful only when the differences between the mean and the values of the distribution are of concern. All the values in the distribution are used to calculate this measure. However, it is not suitable for further statistical analysis.

## 2.8   Measures of Dispersion of Grouped Data

Data is frequently presented in summary forms such as frequency distribution.

## Range

Estimation of the range from grouped data is based upon an assumption that the boundaries of the classes represent actual elements in the data set. If this is so, the range will be the difference between the upper boundary of the highest class and the lower boundary of the lowest class. This estimate will never be less than the actual range. It is quite frequently greater.

## Standard Deviation

The standard deviation can also be estimated from a frequency distribution as shown in Equation 2.7. However, care must be exercised in the adaptation of formulas for calculating the standard deviation. In short, accuracy is sacrificed when one variable in Equation 2.4 is an estimate.

$$\sigma = \sqrt{\frac{\sum [f(CM)]^2 - \frac{\left[\sum f(CM)^2\right]}{N}}{N}} \tag{2.7}$$

where CM is the center mean and f the frequency.

A correction factor is required when an "estimate" is used in the equation. As an example, consider the following data:

The following are the costs in dollars incurred by a company in treating minor accidents during one week: 3.80, 6.38, 8.21, 10.25, 12.90, 13.29, 14.73, 14.76, 15.51, 18.33, 19.57, 20.53, 20.95, 21.70, 23.18, 24.02, 25.06, 31.15, 33.40, 35.62, 38.22, 45.35, 49.43, and 58.85.

The data is summarized and presented in Table 2.1. Using Equation 2.4, the standard deviation is calculated to be ±13.70, Equation 2.5 to be ±13.68 and Equation 2.6 to be ±13.01

**TABLE 2.1**

Computation of the Estimated Standard
Deviation of Example

| Class ($) | f | CM | CM² | f * CM | f * CM² |
|---|---|---|---|---|---|
| 0.00–9.99 | 3 | 5 | 25 | 15 | 75 |
| 10.00–19.99 | 8 | 15 | 225 | 120 | 1800 |
| 20.00–29.99 | 6 | 25 | 625 | 150 | 3750 |
| 30.00–39.99 | 4 | 35 | 1225 | 140 | 4900 |
| 40.00–49.99 | 2 | 45 | 2025 | 90 | 4050 |
| 50.00–59.99 | 1 | 55 | 3025 | 55 | 3025 |

The estimate of the standard deviation in this example is smaller than the true standard deviation computed with the actual values. While some difference between the true and estimated standard deviation can be expected, larger populations will generally result in smaller differences.

## 2.9   Other Measures

The median is a measure that divides a data set into two equal parts. Other measures divide data sets into more than two equal parts such as quartiles, deciles, and percentiles.

## Quartiles

Quartiles, as the name implies, divide a data set into four equal parts. The first quartile, Q1, separates the first and second parts; the second quartile, Q2, separates the second and third parts; and the third quartile, Q3, separates the third and fourth parts. The second quartile is, of course, identical to the median and can be found at the $[(N + 1)/2]$th position in an array. The first quartile can be thought of as the median of the lower half of an array and is at the $[(N + 1)/4]$th position. The third quartile has a similar position in the upper half of an array, the $[3(N + 1)/4]$th position. Quartiles can also be computed from a frequency distribution or grouped data with slight modifications to the formula for the computational median, as shown in Equations 2.8 to 2.10.

$$Q_1 = L_1 + CI \frac{\left(\dfrac{N}{4} - a_1\right)}{b_1} \qquad (2.8)$$

$$Q_2 = L_2 + CI \frac{\left(\dfrac{2N}{4} - a_2\right)}{b_2} \qquad (2.9)$$

$$Q_3 = L_3 + CI \frac{\left(\dfrac{3N}{4} - a_3\right)}{b_3} \qquad (2.10)$$

where L1, L2, L3 are the lower limits of the classes containing Q1, Q2, and Q3, CI is the class interval, and a1, a2, a3 and b1, b2, b3 are analogous to a and b in the formula for the computational median.

## Deciles and Percentiles

Deciles divide a data set into 10 equal parts, and percentiles divide a set into 100 equal parts. Again, there is in each case one less measure than the number of parts, so that deciles will number D1 to D9 and percentiles P1 to P99. Deciles and percentiles are appropriate for larger populations that we have been using in this chapter, but their interpretations are quite similar to the other equal-part measures. For example, the fifth decile D5 and the fiftieth percentile P50 are both equal to the median. The sixth decile D6 marks the point between the lower six-tenths and upper four-tenths of the items in a data set, the 83d percentile P83 separates the lower 83 percent from the upper 17 percent, and so on.

Percentiles are particularly convenient in comparing a single data element with the population or data set. It is, for example, more informative to say that a safety professionals salary at the 74th percentile is $45,000 than to say the median salary is $40,000. Percentiles are also commonly used in describing performance on standardized tests, such as the Scholastic Aptitude Test (SAT) and the Graduate Record Examination(GRE), where a table for converting a standardized test score into a percentile is furnished along with the test results.

## Coefficient of Variation

This is a measure of how much bigger is the standard deviation compared with the mean. Mathematically expressed as shown in Equation 2.11.

$$\text{Coefficient of variation} \frac{\text{standard deviation}}{\text{mean}} 100 \qquad (2.11)$$

## 2.10 Application of Mean and Standard Deviation

The mean is most widely used measure of location and the standard deviation is the most widely used measure of dispersion. Using the mean and standard deviation, we can decide the relative location of any data value. Associated with each data value is the z-score, given by Equation 2.12. The z-score is a standardized value for the item that can be interpreted as the number of standard deviations the value is from the mean value.

$$z = \frac{x - \bar{x}}{s} \qquad (2.12)$$

The standard deviation and mean are used in Chebyshev's theorem, which states that at least $(1 - 1/k^2)$ of the items in any data set must be within $k$ standard deviations from the mean, where $k$ is any value greater than 1. Consider the example when $k = 2$, we can say at least 75% of the items are within 2 standard deviations from the mean. The standard deviation and mean can also be used to identify extreme values in the set of items. The "box plot" is a recent development of presenting graphical summaries of data. It uses measures of location and dispersion. This may be applicable when it is fully developed and accepted.

## 2.11 Rates

Rates are commonly used measures of EHS professions. Examples of different types of rates are presented in this section.

### ANSI

Before the introduction of OSHA, ANSI standards for recording and measuring injuries were adopted by most industries. The ANSI system uses frequency and severity rates that pertain to death and disabling injuries that involved lost time. The rates are based on a schedule of charges for deaths, permanent total, permanent partial disabilities, plus the total days of disability for all temporary total disabilities. These measures relate to the relative frequency of occurrence of major injuries and days lost or charged. For accidents resulting in deaths, permanent total, and permanent partial disabilities, the days charged are determined by the American Standards Scale of Time Charges. The Disabling–Injury Frequency Rate (DIFR), the Disabling Injury Severity Rate (DISR), and Average Days Charged (ADC) are defined as follows:

DIFR is the number of disabling injuries (including illness) per million employee hours worked expressed mathematically in Equation 2.13

$$DIFR = \frac{(no.\,of\,disabling\,injuries)(10^6)}{(no.\,of\,employee\,hours\,worked)} \qquad (2.13)$$

DISR is the number of days lost or charged per million employee hours worked. Days lost include all scheduled charges for all deaths, permanent total, and permanent partial disabilities, plus the total days of disability from all temporary total injuries that occur during the period covered. Expressed mathematically in Equation 2.14

$$DISR = \frac{(total\,days\,charged)(10^6)}{(no.\,employee\,hours)} \qquad (2.14)$$

ADC is the average length of disability per disability injury. Mathematically expressed as Equation 2.15

$$ADC = \frac{No.\,of\,days\,lost/charged}{No.\,of\,disabling\,injuries} \qquad (2.15)$$

## Example

Company TEN employed 500 full-time workers and 200 half-time workers. Table 2.2. is the record of injuries and illnesses experienced by the workers during that year.

### TABLE 2.2

Injuries and Illnesses Experienced by Workers of Company TEN

| Type of Injury or Illness | No. of Injuries or Illnesses | Days Lost or Charged |
|---|---|---|
| Fractures | 6 | 75 |
| Foot lacerations | 10 | 100 |
| Hand lacerations | 14 | 150 |
| Dermatitis | 7 | 250 |
| Total | 38 | 575 |

Using the data given, DIFR = 31.67, DISR = 487.5 and ADC = 15.39.

## OSHA

The OSHA incident rate (IR) is defined by:

$$IR = \frac{\text{No. of recordable injuries} \times (200,000)}{\text{No. of employee hourse worked}} \tag{2.16}$$

The number 200,000 in the above formula represents the equivalent of 100 full-time employees at 40 hours per week for 50 weeks.

## Frequency-Severity Indicator

The frequency-severity indicator equals the square root of the frequency rate times the severity rate divided by 1000.

## Cost Measures

Two cost measures are frequently used in industry: cost factor and insurance-loss ratio. The cost factor is the total compensation and medical cost incurred per 1000 worker-hours of exposure as shown in Equation 2.17

$$\text{Cost factor} = \frac{\text{cost insured} * 1000}{\text{total worker} - \text{hours}} \tag{2.17}$$

The insurance loss ratio is the incurred injury cost divided by the insurance premium as shown in Equation 2.18:

$$\text{Loss ratio} = \frac{\text{incurred costs}}{\text{insurance premium}} \qquad (2.18)$$

## Safe-T-Score

The Safe-T-Score is a dimensionless number. It is used to evaluate records. A positive Safe-T-Score suggests a worsening record and a negative Safe-T-Score suggests an improved record over the past. The limits are +2 and –2 for significantly differences. The Safe-T-Score is based on a statistical quality control test discussed in Chapter 6. It is a test of significance. The formula used is given in Equation 2.19:

$$\text{Safe - T - Score} = \frac{\text{frequency rate now} - \text{frequency rate past}}{\sqrt{\dfrac{\text{frequency rate past}}{\text{million worker} - \text{hours now}}}} \qquad (2.19)$$

## 2.12 Limitations of Rates

Some limitations of the existing rate measures are as follows:

- Rates are not sensitive enough to serve as accurate indicators
- The smaller the sample size, the less reliable are the rates
- A single sever injury or death will drastically alter the severity rate in smaller organizations
- Rates do not reflect environments involving nonparallel hazard categories
- Rates are based on after-the-fact appraisals of injury-producing accidents

## 2.13 Errors in Measures

The art of statistics involves collecting numerical data in a meaningful way. There are traps and pitfalls that surround the investigator's collection and interpretation of numerical data. No doubt some of these errors are obvious while others are subtle. The following are a few errors commonly made by inexperienced investigators.

## Rates

Much numerical evidence centers on the determination of rates, risks, or chances. These measures consist of both a numerator and a denominator. Usually, numerators are easily calculated: however, obtaining the proper denominator may be a challenge and subjected to wrong values. Another error associated with the use of rates is the interpreting of the numerator as a rate.

## Comparison

Much of EHSs measures are comparative exercises. Until recently, most EHS studies were retrospective comparisons. The tendencies in these studies were to draw conclusions without reference to a control group. For example, there was a speculation that workers of nuclear power plants were exposed to low dosages of ionizing radiation. If the exposure were for more than five years, the worker would experience a reduced longevity. To test this speculation, an investigator collected data on the age at death of 3120 nuclear workers who died during the last thirty years, as shown in Table 2.3. Of the collected data, 2184 workers were employed for more than five years and 936 were employed for less than five years.

### TABLE 2.3

Age at Death Distribution of Nuclear Workers

| Age at Death (Years) | % Death per Exposure Group | |
|:---:|:---:|:---:|
| | <5 Years | >5 Years |
| <50 | 21.4 | 13.3 |
| 50 to 60 | 38.6 | 10.0 |
| >60 to 70 | 25.7 | 20.0 |
| >70 to 80 | 21.4 | 40.0 |
| >80 | 7.1 | 16.7 |

The investigator cannot refute or accept the speculation, since there was no control group. The error is to draw conclusions such as fewer workers who were exposed greater than 5 years lived beyond 80 years.

## Normal Distributions

The applications of normal limits are often not valid. Many distributions of EHS do not follow or even approximate to the normal distribution. The error is assuming that the distribution is normal without presenting convincing evidence.

# 3

## Probability and Probability Distributions

### 3.1   Introduction

Probability is merely a way of expressing the likelihood of an event. Qualitatively, it can be expressed as an adjective, for example, "remote" and "quantitatively" as a numerical value. Probability plays an important role in the theory of statistics, therefore, its inclusion in this text book. Probability plays a role in almost in everything we do: for example, mortality rates, weather forecasting, and completion of a project. Events may be simple, that is, able to calculate its probability very easily, or complex, that is, dependent on the probabilities of other events. For simple events, the probabilities may be derived logically, observed empirically, or assigned subjectively. An activity can be a single event or a combination of many events. An event is the possible outcome that may occur.

### 3.2   Simple Events

The fundamental law governing a simple event states that for $N$ different outcomes all equally likely, the probability of $x$ been favorable outcomes are given by $x/N$. This can be written mathematically as $P(A) = x/N$ where $A$ describes the event. Simply stated, the probability of an event $A$ can be stated as:

$$P(A) = \frac{\text{favorable ways an event can occur}}{\text{total ways all outcomes can occur}}$$

Although the idea of probability is the same for logical and empirical processes, empirical probabilities can be understood better if the rules for logical probabilities are modified slightly. The empirical probability of an event is the ratio of the observed frequency of that event to the total number of observations, written as:

$$P(\text{event}) = \frac{\text{observed frequency}}{\text{total frequency}}$$

A third type of simple event probability is subjective probability. This occurs when a process is not readily predictable and observed data are unavailable. Usually a value is subjectively assigned. This assignment may be based on quality factors, experience in similar situations and even intuition. For example, the weather does not follow a known pattern, as result the meteorologist considers the wind directions, pressure areas, adjacent weather system and theoretical knowledge of cause and effect relationships in meteorology to announce an estimate of the probability of rain or snow. EHS personnel are cautioned not to use this type of subjective probability. Whatever the type of probability used to decide a value, if the probability of success is denoted by $p$ and that of failure by $q$, then $p + q$ must equal one, or 100%. Simple events are independent.

As an example of the above concepts, consider a deck of 52 playing cards that are well shuffled. The probability of selecting a spade is:

Number of outcomes                     = 52
Number of spades                       = 13 (favorable outcomes)
Possibilities of drawing 1 spade   = 13/52    = 1/4

Written in mathematical form:

Event $A$ is drawing a spade from the deck
$P(A) = 13/52$

The probability of not drawing a spade is clearly equal to one minus the probability of drawing a spade, written as $P(A') = 1 - P(A)$. Alternatively, it can be viewed as the probability of drawing any one of the other 39 cards.

The probability of 'n' simple independent events all occurring is given by the product of the probabilities and mathematically written as, $P(A) *$ $P(B) * \dots * P(N)$. While the probability of event $A$ or $B$ or $C$ or … $N$ will occur is given by the sum of the probabilities and written as $P(A) + P(B) + P(C) + \dots P(N)$.

## 3.3   Compound Events

When the probability of an event occurring is influenced or related to the probability of another event, the event is termed "compound." These events are either mutually exclusive or conditional. Two events are mutually exclusive when one of them cannot occur if the other occurs. The additional rule of independent probability is applicable here. A conditional event is one that may occur, only if another event has already occurred. The joint probabilities that are all the events occurring are given by the product rule given that another has occurred. For example, two events $A$ and $B$, $P(A$ and $B) = P(A) * P(B/A)$ where $P(B/A)$ reads probability of $B$ given that $A$ has occurred, or alternatively written as $P(A$ and $B) = P(B) * P(A/B)$. The probability that one event will occur, for example $P(A$ or $B)$ is given by $P(A) + P(B) - P(A$ and $B)$. The word "or" is used in the inclusive sense that is either $A$ or $B$ or both.

The idea of probability is used extensively in EHS. As an example, a manufacturer tested a new chemical for biodegradation in his laboratory resulting with a 90% success. For the chemical to be approved by EPA it must also be tested by an independent laboratory and the overall success must be greater than or equal to 75%. The major concern of the manufacturer is obtaining EPA's acceptance of the chemical. If we let activity $A$ be the manufacturer's laboratory test, and activity $B$ be the independent laboratory test, then:

$$P(\text{overall success}) = P(A) * P(B)$$

$$0.75 = 0.90 * P(B)$$

$$\therefore P(B) = 0.833$$

If the national success rate is 40% when not tested in the manufacturer's laboratory and 44% when tested successfully in the manufacturer's laboratory, what is the probability that the independent lab success will result in success in the manufacturer's laboratory? In essence what we are trying to find is $P(B/A)$. We know:

$$P(A) = 0.90$$

$$P(B) = 0.40$$

$$P(B/A) = 0.44$$

Recall $P(B) * P(A \backslash B) = P(A) * P(B \backslash A)$

$$\therefore P(A/B) = \frac{P(A) * P(B/A)}{P(B)}$$

so substituting

$$= \frac{0.90 * 0.44}{0.40}$$

$$= 0.99 = 99\%$$

## 3.4　Expectations

An expected value (EV) is a weighted average value, based on probabilities. This is very useful in decision making. Two common methods using EV to present various decision options and their possible outcomes are in the forms of pay-off table and the decision tree. A pay-off table is simply a table that lists the various options, possible outcomes, and results. Consider two production options recommended for reducing the number of cuts. From these options, a profit profile was constructed as follows:

Option A

| Profit Probability | Profit |
|---|---|
| 0.8 | 5 |
| 0.2 | 6 |

EV of profit = (0.8 * 5) + (0.2 * 6) = $ 5.2

Option B

| Profit Probability | Profit |
|---|---|
| 0.1 | −2 |
| 0.2 | 5 |
| 0.6 | 7 |
| 0.1 | 8 |

EV of profit = (0.1 * −2) + (0.2 * 5) + (0.6 * 7) + (0.1 * 8) = $5.8

Clearly Option B has a higher EV of profit. This is a simple example. A complex example is presented in Chapter 12 "Case Examples."

A "decision tree" is a diagram that illustrates the choices and possible outcomes of a decision. A simple example is, a company can purchase a new machine with the expected production of 15,000 with a probability of 0.8, or repair the old machine resulting in production of 8000 with a probability of 0.9, or continue producing 6000. The cost to replace the machinery is $56,000, and to repair them is $8000. The profit on each item is $1.00. If the difference between the capital cost and expected profit is the only decision factor, then a simple tree can be constructed as shown in Figure 3.1. The triangle is a decision point. At the three nodes A, B, and C, a

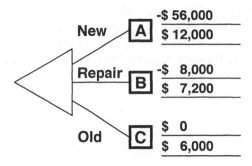

Figure 3.1   Example of a Decision Tree.

common deciding criteria must be established using the factors considered, for example the ending balance, which will be –$44,000, –$800, and +$6000 respectively. Occasionally, a decision taken will influence another decision that might then have to be taken at some time in the future, depending on how the results turn out. When this situation arises, the decision tree can be drawn as a two-stage tree.

## 3.5   Probability Distributions

Probability distribution describes how the probabilities are distributed over the values of random variables. So far we have expressed probability as discrete values. Probabilities can also be expressed as a distribution function, either discrete or continuous. The Binomial, Poisson, and Hypergeometric probability distributions are three discrete distributions that describe many common physical phenomena.

### Binomial

For a probability distribution to be classified as 'binomial' it must satisfy the following four properties:

1.  Each event must be capable or can be defined as having two and only two mutually exclusive outcomes
2.  The probability of an outcome must be the same for every event
3.  Events must be independent of each other
4.  There must be a discrete number of events

Now suppose we have $r$ successes in $n$ independent and similar events, for example throwing a six-sided dice, the probability of throwing six is

1/6. The probability of throwing six, ten times is $1/6 \times 1/6 \times 1/6... =$ $(1/6)^{10}$. This is the product rule discussed previously.

Overall, if there are "$n$" trials and $p$ and $q$ are the chances of success and failure respectively for each trial, the probability of $r$ successes and $(n - r)$ failure despite the order is given by:

$$C_r^n * p^r * q^{n-r} \qquad (3.1)$$

where $C_r^n = n!/[r! * (n - r)!]$ noting that $n!$ is factorial $n$ and equal to the product of $n, n - 1, n - 2, ...$ up to 1.

A plot of probabilities against all possible values of $n$ is a binomial distribution curve. This curve is not unique. A family of curves can result with the shape of each curve depending on the values of $n$ and $p$. For example when $p = q$, the curve is symmetrical about the median no matter the value of $n$. As an example: in reviewing the proposal to automate a panel of four switches, the safety engineer examined the safety records of operating the switches over the last five years. He calculated that the probability of an operator pushing an incorrect switch in the night shift is 0.25 or 25%. Since there are usually three trained operators, he examined the different probabilities of none, one, two or three depressing an incorrect switch during the same night shift. The results are as shown in the Table 3.1

**TABLE 3.1**

Probability of Operators Depressing an
Incorrect Key

| No. of Operators | $f(x)$ | Prob. % |
|:---:|:---:|:---:|
| 0 | $^3C_0 * 0.25^0 * 0.75^3$ | 32 |
| 1 | $^3C_0 * 0.25^1 * 0.75^2$ | 42 |
| 2 | $^3C_0 * 0.25^2 * 0.75^1$ | 14 |
| 3 | $^3C_0 * 0.25^3 * 0.75^0$ | 10 |

As result of reviewing these probabilities, the safety engineer recommended that there should be at least four trained operators per night shift instead of automation. This will reduce the risk to less than 5%. This will result in no capital expenses or job losses, which may be important to management.

These calculations can be laborious. Fortunately, we have tables, programmable calculators, and computers for calculating binomial probabilities. The mean of a probability distribution is the product of $n$ and $p$. The variance is the product of $n, p,$ and $q$. The mean of a probability distribution is called the expected value or expectation of the variable $x$, and is denoted by $E(x)$.

## Hyper-Geometric

The hyper-geometric probability distribution is closely related to the binomial probability distribution. The major difference between the two distributions is that with hyper-geometric the events are dependent. Thus, the probability of success changes from event to event. One of most important applications of the hyper-geometric probability distribution involves sampling without replacement from a finite population. The function is defined by:

$$f(x) = \frac{C_r^n * C_{n-x}^{N-r}}{C_n^N} \tag{3.2}$$

where $n$ is the number of trials, $N$ is the number of elements in the population, and $r$ is the number of elements in the population labeled success.

An example of this type of distribution is a committee comprising of six men and four women in which the committee must select two persons to represent it at a conference. The chance of both people selected are women are equal to $4/10 * 3/9 = 2/15$.

## Poisson Distribution

The Poisson distribution is another form of the binomial distribution. Here, $n$ is large, $p$ is very small and the product of $n$ and $p$ is about unity. A family of distribution curves can also result. The function of the distribution is given by $f(x) = (\mu^x e^{-\mu})/x!$ where $\mu$ is the expected value or average number of occurrences in an interval. There are two assumptions:

1. The probability of an occurrence is a constant for any two intervals of equal length.
2. Occurrences within intervals are independent.

This distribution may arise in any situation where an event is liable to occur at random or at irregular intervals of time or space. As an example when we spread blood on a plate to count the number of organisms on a square centimeter we produce a spread with an occasional flaw. We can conclude that the plate is either with flaws or not with flaws. If there are flaws, we can never accurately determine a probability of a flaw, since the flaw population is an unknown. We can only determine a mean number of flaws per unit of area. Here, the Poisson distribution should be used instead of the Binomial distribution.

## 3.6 Continuous Probability Distributions

Three continuous probability distributions will be discussed under this heading: normal, uniform, and exponential.

### Normal Distribution

The Normal Distribution is perhaps the most frequently used distribution used to describe a continuous random variable. It is an ideal curve that nearly fits many distributions in statistics. For the benefit of the mathematically minded person, taking the mean as zero, the function is written as:

$$f(x) = \frac{1}{\sqrt{2\Pi}\sigma} \exp^{-}(x-\mu)^2 / 2\sigma^2 \qquad (3.3)$$

where $\sigma$ = standard deviation and $\pi$ = 3.142.

The form of the curve is illustrated by a bell shape. We note that the normal distribution is symmetric, the standard deviation determines the width of the curve, the mean can be any numerical value, the total area under the curve is equal to one, and the probability for any normal random variable is given by the area under the curve. The normal distribution with mean ($\mu$) = 0 and $\sigma$ = 1 is called the standardized normal curve. The points on the abscissa are denoted by $z$ and range from +1 to −1. As indicated, the probability for any normal random variable ($x$) is given by the area under the curve. To convert a variable to a standard normal $z$ value we simply use the formula:

$$z = \frac{x-\mu}{\sigma} \qquad (3.4)$$

Once we calculate $z$, standard normal distribution tables are available for determining the probabilities that are equal to the area under the curve. For any normal distribution, the relation above permits the use of the tables of the standardized normal distribution. The quantity $(x-\mu)/\sigma$ is called by many different names such as critical ratio, relative deviate, standardized value, or normal deviate.

### Uniform

This is a constant function over a given measure. For example, consider a machine producing $x$ items per minute and all things being equal, the probability of producing $x$ items in the 5th, 10th, or 15th minute will be the same. This is not frequently used in EHS and business studies.

## Exponential

The continuous exponential distribution is often used to describe the probabilities of occurrences of an event in terms of space and time. Many events follow this type of distribution, such as failures of machine parts due to wear and tear and decay. The function is given by:

$$f(x) = \frac{1}{\mu} e^{-x/\mu} \qquad (3.5)$$

for $x \geq 0$ and $\mu > 0$.

---

## 3.7  Guidelines in Selecting a Probability Distribution

1. **Binomial** — preferred when there are only two possible outcomes and the events are independent.
2. **Poisson** — preferred when the distribution is a binomial with a large sample size and small favorable probabilities.
3. **Hyper geometric** — preferred when there are only two possible outcomes and the events are dependent.
4. **Normal** — preferred when the outcomes are continuous and can display extreme conditions
5. **Exponential** — preferred when the outcomes are continuous and decreasing or increasing with time or space.
6. **Uniform** — preferred when the outcomes are the same in time or space.

---

## 3.8  Examples of the Use of Probability Theory

## Example 1

Company TEN operates two factories at East and West Sutton. At each location, the safety manager wishes to obtain some information about defective items of production which damaged the safety guards. At the East factory, product A is susceptible to two types of defect. There is a 0.15 probability of inspected output having defect L, and a 0.14 probability of it having defect M. The defects occur independently. The factory also manufactures another product, B, which is susceptible to three types of defect: L, M, and N. The probability that product B has defect N depends on whether it contains any other defect, L or M. The probability that an item of product A has the following defects were calculated for:

1. At least one defect
2. Both defects, L and M
3. One defect only
4. No defects at all

and the probability that an item of product B has:

5. None of the three defects
6. One of the three defects.

Given the following for product B:

The probability of having defect L is 0.15

The probability of having defect M is 0.14

The probability of having defect N if it has neither L nor M is 0.3

The probability of having defect N if it has one of L or M, but not both is 0.2

The probability of having defect N if it has both defects L and M is 0.1

At the West factory, product C goes through five consecutive operations before it is completed. The only information about production known and thought to be typical of normal operations, is as follows:

| Starting Process | 7500 Rejects |
| --- | --- |
| 1 | 660 |
| 2 | 520 |
| 3 | 430 |
| 4 | 290 |
| 5 | 80 |

The manager calculated the following probabilities:

7. That a unit, once started, will become a completed unit of finished product
8. That a unit will get beyond the second process

9. If an additional 2000 units of finished product C are required, how many more units should be started

Knowing these probabilities will help the safety manager to determine the risk of the machine guard being damaged, the labor required, and establish a maintenance schedule.

Outline of the calculations are as follows:

1. Probability of a defect, P(L or M)     = P(L) + P(M) - P(L and M)

  Probability of having defects, L and M     = P(L and M) =P(L) * A(M)

                                       = 0.15 * 0.14

                                       = 0.021

  P (L or M)                                = 0.15 + 0.14 - 0.021

                                       = 0.269

2. Probability of having both defects is 0.021, as calculated in Step 1
3. Probability of one defect only is the solution to (1) minus the solution to (2), i.e., 0.269 – 0.021 = 0.248
4. Probability of no defects at all = 0.85 * 0.86 = 0.731
5. We know the probabilities of product B having defects L and M are as follows:

| | | Probability of Defect N |
|---|---|---|
| Both L and M | 0.021 | 0.1 |
| L or M, but not both | 0.248 | 0.2 |
| Neither | 0.731 | 0.3 |
| | 1.000 | |

The probability of having no defect at all is:

P (Neither L nor M, not N) = P (neither L nor M) • A (not N)

                            = 0.731 * 0.7

                            = 0.5117

6. The probability of having one defect only is the sum of:
   P (L or M but not both, not N) plus P (neither L nor M, but defect N)

   The probability of defect L or M, but not both, without defect N

$$= 0.2248 * 0.8$$

$$= 0.1984$$

The probability of defect N, but not L or M

$$= 0.731 * 0.3$$

$$= 0.2193$$

The probability of one defect

$$= 0.1984 + 0.2193$$

$$= 0.4177$$

7. Total number of rejects = 1980
   Total number of successfully completed units

$$= (7500 - 1980)$$

$$= 5520$$

   Probability of successful completion = 5520/7500 = 0.736
8. Rejects in processes 1 and 2 = 1180
   Number successfully getting through to Process 3

$$= (7500 - 1180)$$

$$= 6320$$

   Probability of completing process 2 = 6320/7500 = 0.8427
9. Let the number of units started to complete 2000 finished items
   be x:

$$\text{then } 0.736x = 2000$$

$$\text{therefore } x = 2718$$

## Example 2

Company EIGHT has three chauffeur-driven cars for the use of senior managers. They can reserve a car for an entire day for their own exclusive use. Records showed that nine out of ten times when a manager was self driven, a vehicular accident occurred. The fleet safety manager in planning for the preceding year will like to know the probabilities of a manager driving, hence the risk of an accident occurring. She established that the daily demand for cars follows a Poisson distribution because of the random nature of demand for cars, even though n = 3 and a mean of 0.5. If all three cars are used equally, she calculated the probabilities of:

1. No car being used on any one day
2. A particular car not being used on any one day
3. All cars being used

The calculations are as shown:

$$m = 0.5 \text{ (given), and } e^{-m} = 0.60653$$

1. $P(0) = \dfrac{0.5^0}{0!}(0.60653) = 0.60653$

   Therefore the probability of no car being used is 0.60653.

2. There are three occasions when a particular car may not be used:

   a. No car use
   b. One car but not this one
   c. Two cars but not this one

   no cars used: $P(O) = 0.60653$; one car but not this one: $P(1) = 0.5/1 * (0.60653) = 0.30326$ and probability that this is not the one under consideration $= 2/3 * 0.303265 = 0.20218$; and two cars used, but not this one: $P(2) = (0.5)^2/2 * (0.60653) = 0.07582$ and probability that the car under consideration is not used is $= 1/3 * 0.07582 = 0.02527$

   Therefore $P(0) + P(1) + P(2) = 0.83398$.

3. All the cars will be used if the demand is 3 or more:

$$P(3 \text{ or more}) = 1 - A(0) - A(1) - A(2)$$

$$= 1 - (0.60653 + 0.30326 + 0.07582) = 0.01439$$

The probability that all cars will be used is 0.0144 or less than 1.5%.

*Section B*

# Data Gathering and Analysis

# 4

## Sampling Methods

### 4.1  Introduction

The collection of data, both quantitative and qualitative, from large populations are often impossible, impractical, or too costly. One major objective of collecting data is to extract information that are "truths" or nearly "truths." Is it necessary to examine the entire population to obtain truths? In other words, do we collect all possible data or use all the data we collected? Do we conduct a survey or a census? A census is generally a complete examination of a population, whereas a survey is usually an examination of a sample. We will attempt to answer this question at the end of this section.

Sampling is the selection of part of a population to represent the whole population. The law of statistical regularity states that a set of subjects taken at random from a large group tends to reproduce the characteristics of that large group. This is the law underlying sampling. So what is important about sampling is the selection process. It must be rigorous and reliable. Calling a few handy items a sample, and drawing sweeping conclusions about the population is all too easy. Therefore, "convenient sampling" is the least preferred scientific method of sampling. The validity of the sample depends whether the errors introduced by the sampling process are sufficiently small not to invalidate the results for the purposes for which they are required. Inevitable there will be errors termed "random-sampling" errors. These errors depend on the size of the sample, the variability of the population, the sampling procedure adopted, the data collected and the analysis of the data.

### 4.2  Requirements of a Good Sample

The essence of a good sample is no 'bias' and minimum 'random errors'. Bias occurs whenever the characteristic to be measured is influenced or lie in the same direction. For example, in the monthly reports on the results of random inspection of fast-food outlets for evaluating their hygiene

facilities, the area health inspectors are only required to submit a list of all
the fast-food outlets that are not acceptable and operating. To enhance
their reputation as good health inspectors, the tendency is to select a sample of outlets suspected or known offenders. This will result in a high percentage of outlets classified as offenders. Using this data submitted by the
inspectors to draw conclusions about the population of outlets will always
be 'bias'. There are many ways in which faulty selections of the sample
may cause bias.

1. Deliberate selection of a "representative" sample
2. The selection of sample depending on some characteristic, which
   correlates with a point of interest
3. Conscious or unconscious selection
4. Substitution in difficulties
5. Failure to cover the whole designed sample

Clearly if bias exists, no true conclusions can be drawn from the sample.
Random selection is the best method to avoid bias. Note that random
selection does not mean haphazard selection. There are many ways or
techniques to achieve random selection. As mentioned previously, random errors are a result of the sampling technique. Calculation of random
errors will be discussed using examples.

Let us examine the sample presented in Table 4.1 to decide if it is a true
representation of the population. The sample was collected to learn if there
is any relationship between incidence morbidity rates and the size of
households. The data was collected to analyze incidence morbidity for the
year 1978 based on household size distribution.

**TABLE 4.1**

Morbidity per Household Distribution

| No. in Household | % of Sample | % of Population | Incidence Morbidity Rate |
|---|---|---|---|
| 2 | 19.4 | 26.8 | 205 |
| 3 | 20.9 | 26.5 | 260 |
| 4 | 23.5 | 21.9 | 325 |
| 5 | 15.4 | 13.0 | 300 |
| 6 | 8.1 | 5.9 | 375 |
| 7 | 8.5 | 3.2 | 360 |
| 8 | 1.9 | 1.5 | 400 |
| >9 | 2.2 | 1.3 | 500 |

*Note:* Morbidity rate is per 1000 people.

This sample is not a good representation of the population; therefore, the objective of the survey cannot be truly achieved. Most sampling techniques demand a subdivision of the population into units, termed sampling units, which form the basis of the actual sampling procedure. As mentioned previously, the principal objective of any sampling procedure is to secure a sample that will reproduce the characteristics of the population. It must be understood that the sample size will be influenced by the restrictions of time, money, skilled labor, and other resources.

## 4.3 Sample Size

Selecting a sample size involves a trade-off between precision and cost. Larger samples provide greater precision and cost. A common approach to deciding sample size is to first specify the precision required and then select the smallest possible sample size. The sample size can be determined using Equation 4.1, where $B$ is the boundary or precision of our estimate; for example, 95% confident, $N$ the population, $n$ the sample and $s$ the standard deviation of the sample.

$$n = \frac{Ns^2}{N(B^2/4) + s^2} \qquad (4.1)$$

The standard deviation can be determined by using the approximate formula of largest minus smallest value in the data set divided by 4. If this method is not suitable there are several ways suggested in test books on sampling techniques.

## 4.4 Sampling Systems

### Random
A Random sample is a sample where each member of the population has the same chance of selection as any other, and there is no connection between the chances and selection of different members. Random sampling is the standard to which complicated sampling techniques aspire to achieve.

## Stratified

Stratified sampling is useful when the population can readily be divided into many groups or strata, so that each member of the population falls in one and only one stratum. Sampling is taken separately in each stratum and the results are pooled appropriately at the end. Clearly these strata may all contain the same number of elements, or different numbers of elements. If a uniform sampling fraction is used, it is termed stratification with uniform sampling fractions. Stratification has two purposes. The first is to increase the accuracy of the overall population estimates and the second is to ensure that subdivisions of the population that are themselves of interest are adequately represented. Stratification with a variable sample fraction is sometimes necessary to improve the reliability. Considerable gains in accuracy can result if the sampling fractions ($f$) are proportional to the within-strata standard deviations ($\sigma$) of the members, that is,

$$\frac{f^1}{\sigma_1} = \frac{f_2}{\sigma_2}$$

A population may be stratified for two or more different characteristics, termed multiple stratification; for example, the classification of chemicals according to its percentage of lethal ingredient and flammability. The sample size for each stratum, when estimating the population mean can be determined using Neyman's allocation (Equation 4.2)

$$n = \frac{\left(\sum_1^s N_s * s_s\right)^2}{\sum_1^s N_s * s^2 + N^2 \frac{(B^2)}{4}} \tag{4.2}$$

where $n$ is the sample size of the strata, $N$ is the population in the strata, and $B$ the precision boundary.

## Systematic

Systematic sampling or quasi-random sampling consists of taking every mth element from subsets. The 'mth' element is chosen randomly from the first subset. The number of elements in a subset is determined by the sample population. A systematic sample would be equivalent to a fully random sample if the list were arranged wholly at random. One should be careful in using this technique. For example, selecting the tenth name from a list consisting of couples of which the males are listed first then the females. Here, the samples would be entirely females. If the sampling is system of selecting the mth member is random, then sample size can be determined as random sampling.

## Multistage

In multistage sampling the population is regarded as made up of many first-stage sampling units, each of which is made up of many second stage units and so on. At the first stage, the first-stage units are selected by a suitable method, such as random or stratified sampling. At the second stage, samples from the first stage selection are selected again by some acceptable method to form the second stage sample. Further stages may be repeated as required. This method is less accurate than taking a simple random sample of the same size. The usefulness of this method is the identification of the sample based on established criteria.

## Multiphase

Multiphase sampling is choosing samples from a sample to form sub-samples. Note that multiphase sampling differs structurally from multistage sampling, in that multiphase sampling requires the same elements, whereas in multistage sampling a hierarchy of sampling elements is used.

## Quota

Quota sampling is a non-random technique. This technique demands a predetermined number of elements based on some criteria to form the sample. For example, interviewing the first hundred people waiting to enter a concert hall. This method must be used with caution.

## Public Opinion

Public opinion polls should be interpreted with caution. They are not recommended to be used in EHS for collecting data to establish truths.

We have now briefly discussed sampling methods. The aim of sampling is to obtain a sample that would reflect the characteristics of the population to be measured. The challenge now is to decide the size of the sample, the sampling method, the desired accuracy, and the reliability of our measured sample values. The reliability, accuracy, and validity should be clearly shown in our objective statements. As previously mentioned, the sampling technique should be decided on based on our objectives for collecting the data. For example, if our objective is to obtain general opinions from the public at large to raise awareness, then quota or opinion polls may be the appropriate sampling method.

## 4.5  Sampling Considerations

The practical problems encountered in planning to sample vary greatly with the type of population and the nature of the data to be collected. There are however, a few common points to be considered:

1. Specification of the objectives
2. Definition of the population
3. Decision on the nature of the information to be collected.
4. Decision on the method of collecting the data
5. Choice of a frame
6. Choice of sampling unit and type of sample
7. Decision on the publicity

These points are interrelated and cannot be considered individually. For example, when defining the population, the choice of sampling unit and type of sample will influence the decision in cases where there are marginal categories and separate units are not desirable. The data to be collected varies from subject to subject, purpose to purpose, and so on. What is important is that we set out to collect the types of observations that have a bearing on the subject under investigation. Data to be collected must be appropriately selected so that a coherent whole is obtained that adequately covers the subject of investigation. When this is not possible, the objectives of the investigation must be reexamined and modified. Occasionally, working in collaboration with experts on the subject is essential.

The practice of collecting excessive data may be a good idea. Unfortunately, this is neither an economical nor a desirable proposition. The data to be collected should be useful and necessary. The tendency of developing questionnaires or formulating questions as the means of collecting data must be guarded. Observations are preferable to questions. Questions on facts and past actions are preferred to questions on generalities and hypothetical future outcomes. Inspecting an area for cleanliness is better than to ask questions such as "Do you have empty containers on the premises?" There are so many things to be considered in asking questions such as, are the respondents sufficiently informed to answer. As an example, consider this question: "Do you prefer chlorine or fluorine in your drinking water?" Certainly many people cannot answer this question because they are not sufficiently informed. Other concerns involve the wording of the questions and order of the questions. These are all extremely important in developing a questionnaire. The main points to be borne in mind when conducting a questionnaire is presented in the following section. A detailed discussion on developing questionnaires is beyond the scope of this book. Quantifiable observations are preferred. As mentioned, observations are not always quantifiable: however, standardize if possible and calibrate qualitative observations. For example, for a walk-through safety inspection of a plant, a checklist may use a ranking scale of 1 to 5 to show the urgency of correction required for each observed hazard. The selection of a suitable sampling frame is fundamental to a good sample; for example, a list of individuals is not a suitable frame for sampling households unless the individuals are grouped by households.

## 4.6 Developing a Questionnaire

The main points to keep in mind are:

1. Keep it brief as possible
2. Questions should be short
3. Questions should not be phrased to show bias
4. When possible use "yes" or "no" answers
5. Calculations should be avoided
6. Questions should be logical
7. Personal questions should be at the end — optional and generic
8. The purpose of the questionnaire should be explained
9. The questions should be adequate to obtain the required information

## 4.6 Execution of a Survey

Identifying all the problems one is likely to encounter in executing a survey is not possible. Some obvious points are often over looked. There should be a pre-execution stage. In this stage the following subjects should be addressed:

1. **Organization and administration** — This will vary with the level of sophistication and the scale of the survey.
2. **Design of forms** — Careful attention should be given to the design of the various forms that will be used during the survey, especially those on which the observations and answers to questions are to be recorded. As mentioned, the practical ability of obtaining the observations must be considered. Instructions and explanatory notes should be provided where necessary. The latitude to make errors on the form should be diminutive; for example, figures that have to be summed by the field investigators

should be arranged vertically and not horizontally. Testing or conducting a dry-run of the form is advisable.

3. **Selection, training, and supervision of field investigators** — It may be worth the investment to rigorously evaluate all the prospective investigators doing the survey tasks. By evaluating these investigators, differences in collecting the data may be revealed.

4. **Control of accuracy of the field workers** — The best assurance is to ensure that the field investigators are thoroughly trained, capable, and conscientious. Supervision, check and balance points, and training are the ingredients to accurate work.

5. **Arrangements for non-response** — Follow-up arrangements for non-response must be properly made and executed. Note that unplanned substitution for non-response is a dangerous practice.

6. **Collecting, collating, coding, and processing of data** — The use of the computer is normal and there are many statistical software packages on the market. Collating and analyzing data can be a data-processing function requiring specialized knowledge. Remember that the computer only reproduces what is entered: "garbage in, equals garbage out."

7. **Statistical analysis** — The attainment of a high standard of accuracy in computations demands careful organization of the coding and data. Whatever the method of calculation or the statistic calculated, the results of every analysis should be carefully scrutinized for apparent inconsistencies and irregularities. Any anomalous values should be thoroughly investigated and adjusted and compensated for deficits in the sampling procedure. For example, at the end of a housing survey it was discovered that a class of the population was incorrectly represented in the final sample. In such a case where the error was clearly identified, adjusting the results by weighing the different classes enough to compensate for the errors in the proportions was possible. Adjustments should not be made for improper sampling or defects in the sample, in such a case, reporting the results is better as calculated or obtained.

It is advisable to clearly identify the types of statistical inference that can be made with certainty from those that will be speculative. For example, in a nutrition survey it was found that children in families greater than five people were more deficient in calcium than children in families of less than five. We can draw the conclusion that children in families greater than five people are at greater risk of calcium deficiency. We can probably conclude quantitative estimates of the degrees of deficiencies that actually existing among children of families of different sizes. We cannot infer with certainty that the size of a family is the cause of calcium deficiency. Be very careful, that even in situations where a definite causal relationship seems apparent, that deductions of the magnitude of the relationship are not automatic. To decide with certainty the magnitude in the causal sense of the effect of any given factor, further experiments must be conducted. These experiments can be used to decide with a high degree of certainty, if and only if the effects of extraneous factors are eliminated. For example, in the nutrition survey there were many extraneous factors such as water quality, availability of foodstuffs rich in calcium, life style, and so on. These factors can have a direct consequence on the nutrition of children. Review the following data on illness due to a rash in elementary schools as shown in Table 4.2. What conclusions can be drawn from the data presented? With what certainty can you say that areas with many children have most cases?

**TABLE 4.2**

Incidence of Rashes in Elementary School Students, 1994

| Area | No. of Schools | Total No. Children | No. of Cases |
|------|---------------|--------------------|--------------|
| East | 174 | 18,325 | 1482 |
| West | 177 | 17,781 | 1425 |
| North | 189 | 16,891 | 1415 |
| South | 182 | 15,652 | 1324 |
| Central | 179 | 14,367 | 1368 |

There are different ways certainty, and relationships of data collected can be discussed. Discussion on certainty and relationships are covered in Chapter 7. In short, your conclusions will be limited by the raw data you collected.

8. **Reporting** — When the analysis of an investigation is completed, compiling the results in a report is necessary. In addition to the presentation of the numerical results as tables or graphs, some discussion and interpretation of the results are necessary. Generally a good report will include, a general description of the survey, its purpose, a description of the material covered, the nature of the information collected, the methods used in collecting the information, the sampling method (repetition, non-response, point of time), the date and duration, the method of selecting sampling-units, the personnel and equipment, the cost, the accuracy of the survey, the degree of agreement among the investigators, a comparison with other sources of information, and a list of references.

## 4.8   Survey Errors

Two types of errors can occur when conducting a survey. These are sampling and non-sampling errors. Sampling error is the difference between the sample estimator and the population parameter. Non-sampling errors are those due to execution of the survey such as the measurements, non-response, and processing. Non-sampling errors can be reduced by careful planing.

## 4.9   Recommended Attributes of Measurements

When deciding on the sampling technique, it is essential that you carefully define your measures. The following are seven considerations of a good measure:

1. It is quantifiable
2. It use an acceptable and consistent unit of measure
3. It is capable of duplication
4. It can be validated
5. It can generate error-free results
6. It is administrative feasible
7. It is efficient and understandable

The above seven considerations show that selecting a suitable sampling technique is complex. It is unlikely to find all the desired considerations in a single measure, however, every effort should be directed towards selecting an acceptable measures.

## 4.10   Design of Experimental Studies

EHS studies are either experimental or observational. In experimental studies one or more variable is controlled, while in observational studies no attempt is made to control variables. In both cases, data is collected. In experimental design, the variable of interest is called the dependent variable, while in observational design it is called the observation. Also note that observations are not usually divided into independent or dependent. The results of a visual inspection are perhaps the most common type of observation. As an example: it is suspected that the number of workers involved in skin-burn accidents in the sorting and packaging room of the manufacturing operation is related to the failure of one hydraulic seal in the system. An observational study will record the number of workers burned, or the time of occurrence, or the number of hours the seal is in operation before it fails. On the other hand, an experimental study will postulate a hypothesis or question, such as "Is there a relationship among the failures of the seals, number of workers burned, and the operating pressure of the system?" or "the uncertainty of the failure of the seals is the cause of burns received by workers." The question for example, will lead into a series of sub-questions for developing hypotheses, such as "what is the chance of a seal failing and a worker been burned?" In experimental design terminology, the selected units or workers are called experimental units, while in observational design they are called elements. Unlike observational studies, complete randomization and replicability are two important principles necessary in experimental design.

Generally, in designing an experimental study, the problems or questions to be resolved or investigated are clearly stated as a statement. Based on this statement, subproblems and hypothesis are developed. The data needed to conduct the study is determined by the test of these hypotheses. How much data is to be collected and treated is determined by the validity and reliability of the desired results. Studies must be properly planned if meaningful information is required. The first step is to answer the following four simple questions affirmatively. This will confirm that the study is valid.

1. Is the problem statement clearly stated and unbiased?
2. Is the study hypotheses realistic and measurable?
3. Is the focus of the study wide to result in meaningful conclusions?
4. Would such a study be supported by those of concern such as management?

## Presentation of Proposal

Following are some basic guidelines for presenting your proposal to conduct a study:

1. The problem statement or question.
   - Express the thoughts fully with the least words possible, and in short, succinct sentences.
   - Look critically at each thought as it stands on the paper. Do the words say exactly what you want them to say? Read carefully, phrase by phrase, to see if one word can carry the

burden of two or more. Throw out superfluous and unnecessary words.

- Develop subproblems by reading the problem statement critically to discover the areas that should receive in-depth treatment before the problem can be resolved. Every subproblem should contain a word or phrase that shows the necessity to interpret the data within that particular subproblem.

2. The literature review.

- Review pertinent data and literature. This may reveal studies similar to your own, and show how collateral researchers handled these situations. The literature review may suggest avenues of approach to the problem, or reveal sources of data that you may not have known existed, or give you new ideas and approaches, or help you in evaluating your own research efforts by comparing them with related efforts done by others.

- Show the relatedness of the literature to the research project.

3. Think openly.

- We perhaps err in thinking too narrowly by restricting the problem to one belief or perception. Initially, think of the problem in broad generic terms or areas of people, things, records, thoughts and ideas, and dynamics or energy. For example, the failure of the seals may be related to the handling of the control lever or the design of the lever. It may not be related to the pressure in the system. Clearly, we will err if we only focused on the pressure.

4. The study plan.

- Do not confuse a study plan with study methodology. In planning the study project, considering the nature of the data that the investigation or study demands is extremely important, as is the availability, collection, and interpretation of the data. For example: it is postulated that the workers are burned out because of slow-reflect action. Can you research this statement in a work environment without disrupting production? This is only possible if the workers and management are willing to cooperate in the study, and the technical knowledge to measure reflect action and the ability to interpret the results of the tests are at hand. Giving up such a project in favor of one in which you have the knowledge, the resources, and the skill to bring it to completion is probably better. Note that careful studies should be practical, built upon clear and realistic planning, and should be executable within a given framework of a clearly, and feasible design.

5. Methodology.

- Methodology is merely an operational framework within which the data are placed so that their meaning may be seen more clearly. A review of some standard research textbooks will outline a broad spectrum of methodologies.

## 4.11 Quality Control Sampling

Quality control consists of making decisions based on a series of inspections and measurements. There are two sampling methods: acceptance sampling and statistical process control. These two will be briefly discussed. Acceptance sampling is used in situations where a decision has to be made to accept or reject based on an established quality. Statistical process control is used to monitor a production process to learn if it is in control or out of control. Hyper-geometric probability distribution can best describe acceptance sampling. Sampling tables are produced such as the American Military Standard Table.

The control charts can be classified by the type of data they contain. For example, an $\bar{x}$ chart is used for cases where the quality measure is a variable such as length, weight, or time. Two lines are drawn to decide the acceptance region. The control line in this instance is the average. Errors can be either Type I or II. Points outside the control region suggest that the process is out of control. Patterns of data points inside the region can also suggest potential quality control problems.

## 4.12 Considerations in Statistical Analysis of Occupational and Environmental Exposures

Special statistical methods for collecting and analyzing occupational and environmental exposure may be mandatory. Monitoring data usually consist of area samples, personal inhalation samples, biological monitoring, or dermal samples. Area samples are collected to represent the concentration in a specific location over a given period of time. Personal samples are collected to represent a person's exposure during a specified time period. Exposure data collected for each type of exposure should be separated and statistical analyze separately.

Of special interest is biological monitoring, which may also be used to determine exposure. While biological monitoring provides information, interpretation of data can be difficult due to variability in the physiological and health status of the subjects, exposure sources, subject existence, analytical errors, and so on.

Occupational exposure data can be divided into three categories: category one, where all important variables are known; category two where important variables are not known but assumptions can be made for their estimation, and category three, which consist of measurement summaries, anecdotal data, or other data for which the important variables are not known and cannot be estimated. Each category should be treated differently.

## 4.13 Presentation of Report

A report can be divided into five components: introduction, design of investigation, observations, results, and analysis and conclusion. The following outlines the queries to be answered in each component in a typical report.

1. Introduction.
    a. What are the objectives or questions to be answered, and the hypotheses of the investigation?
    b. What is the population of the investigation?
    c. What are the limitations of the investigation?
2. Design of the investigation.
    a. Was the investigation an experiment, planned observations, or an analysis of records?
    b. How was the sample selected? Are there possible sources of bias? If so, what provision was made to deal with this bias?
    c. Is a control group or standard necessary? If so, what is the nature of the control group or standard of comparison?
3. Observations.
    a. Are terms used, measurements made, and criteria of outcome clearly defined?
    b. Was the method of measurement consistent for all the subjects and relevant to the objectives of the investigation?
    c. What are the possible sources of sampling and measuring errors? How were they compensated or treated in the analyis?
    d. Are the observations reliable and reproducible?

4. Results.

   a. Are the findings presented clearly, objectively, and in sufficient detail to enable the reader to draw conclusions for himself?

   b. Are the findings internally consistent such as do the numbers add up properly?

5. Analysis of the results and conclusion.

   a. Are the data worthy of statistical analysis? If so, are the methods of statistical analysis appropriate to the source and nature of the data and is the analysis correctly performed and interpreted?

   b. Is there sufficient analysis to determine the significance of the investigation?

   c. Which conclusions are justified by the findings? Which are not? Are the conclusions relevant to the objectives of the investigation?

# 5

## Sampling Theory and Estimation

### 5.1 Introduction

In Chapter 4, numerical values of the sample were used to estimate the standard error of the statistic of the population. This is a measure of the average error to be expected. The analysis or calculation did not address the frequency with which errors of different magnitudes may occur. Fortunately, most distributions of errors approximate to the normal distribution. As indicated, the larger the sample on which the estimate is based, the closer it will follow the normal distribution. For small-sized samples, with less than 30 elements, the Student t distribution is used instead of the normal distribution. Thus, we use the normal or the Student t distributions to decide standard errors. The sampling and estimation theories discussed in this chapter is most reliable when the population is approximately normal and the sample is selected randomly. It can, however, be used in other situations after making approximations and assumptions but the estimates will be less reliable.

### 5.2 Estimates

We have discussed how to select a good sample. The task is now to calculate an estimate. There two types of estimates: point and interval. Point estimation is a value of a sample statistics that serves as an estimate of the population parameter; for example, the mean $\bar{x}$, and the standard deviation &. Calculation of point estimates was discussed in previous chapters. Interval estimates provide desired precision information. Often the precision information is essential in evaluating and interpreting the sample results. Proportions are treated in a similar manner as a discrete parameter.

Estimation is the simplest form of inferential statistics, which uses known sample evidence to draw conclusions regarding unknown population characteristics. Estimation can be discussed in two groups, point and interval. A point estimate is a numerical value assigned to an unknown population parameter, while the interval estimate is an interval assigned to the population parameter. In statistical investigations, the computed value of a sample statistic serves as the estimate. That statistic is called the estimate of the unknown parameter. Before using an estimate to present or infer conclusions, the test for bias, efficiency, and consistency should be performed. The estimate is not biased when the expected value of the sample is equal to the population parameter being estimated. How can we decide that? This will be discussed in Chapter 6, "Hypothesis and Testing." The efficiency is measured by the magnitude of the variance. The smaller the sample variance the more efficient is the estimate. Consistency is measured by the closeness of the estimates to the population parameter. A consistent estimate improves the reliability and precision.

## 5.3   Point Estimators

### Means

The most frequently used point estimator is the arithmetic mean. If we consider randomly selecting different samples from the same population, different numerical values of the mean will be obtained. The probability distribution of the mean $\bar{x}$ is called the sampling distribution of $\bar{x}$. Knowledge of the sampling distribution and its properties will enable us to make probability statements about how close the sample mean $\bar{x}$ is to the population mean $\mu$. The expected value of the mean denoted by $E(\bar{x})$ is really the mean of the means. Theoretically, this value should equal the population mean. The standard deviation of the mean of the means or the expected value is given by Equation 5.1 for infinite populations where $\sigma$ is the population standard deviation, and $\sigma_{\bar{x}}$ is the standard deviation of all possible $\bar{x}$ values.

$$\sigma_{\bar{x}} = \frac{s}{\sqrt{n}} \tag{5.1}$$

For a finite population, the infinite standard deviation is multiplied by a correction factor as given by Equation 5.2.

$$\text{Finite correction factor} = \sqrt{\frac{N-n}{N-1}} \tag{5.2}$$

Often the population distribution is not known. In these cases we use the central limit theorem to calculate the estimate. The Central Limit Theorem states that the sampling distribution of the sample mean $\bar{x}$ can be approximated by a normal probability distribution as the sample size becomes large. It must be noted that our sample mean will never be exactly equal to the population mean. The absolute value of the difference between the value of the sample mean and the value of the population mean is called the sampling error.

## Proportions

The standard deviation of $p_{avg.}$ are as shown in Equation 5.2 for infinite populations.

$$\sigma_{P_{avg.}} = \sqrt{\frac{p(1-p)}{n}}$$ (5.3)

## Confidence Intervals

There are three methods for determining the end points of a confidence interval of the mean $\mu$ or an other statistical estimate.

1. Using the normal distribution z values when the sample is large as shown in Equation 5.4

$$\mu = X \pm z \frac{s}{\sqrt{n}}$$ (5.4)

2. Using the finite correction factor with the z values for small samples, as shown in Equation 5.5

$$\mu = \bar{x} \pm z \frac{s}{\sqrt{n}} \frac{\sqrt{(N-n)}}{N-1}$$ (5.5)

3. Using critical t value ($t_{\alpha/2}$) from the t-distribution, as shown in Equation 5.6.

$$\mu = X \pm t_{\alpha/2} \frac{s}{\sqrt{n}}$$ (5.6)

## 5.4 Relationship between Mean Point Estimate and Sample Size

As stated, the purpose of sampling is to provide information about a population based on information contained in the sample. Consequently determining the sample size is critical. The size of the sample depends on the distribution of the characteristics in the population; therefore, the population must be clearly defined. Most of the populations in EHS investigations are finite. Therefore, the population correction factor must be applied.

Prior knowledge of our population characteristics can be advantageous to our determination of sampling technique. In cases where little or nothing is known about the population, a pilot or an exploratory survey may be necessary. Even if there is adequate knowledge of the statistical properties of the investigation, a pilot survey is advisable before large-scale investigation. This will also help identify latent practical problems. Pilot investigations can be used to test field procedure and schedules, train field workers, and even identify improvements. Opinion surveys should not be used as pilot investigations. Results from these surveys are doubtful since an individual's opinion on a given subject is frequently both ill-defined and liable to change.

The big question is, how small or large should my sample be? It is believed that the larger the sample size the more accurate the estimates to the actual values. This is true if point estimates are required. When intervals are used, increasing the confidence limits can increase the accuracy or confidence. If our distribution is normal, we can use the z tables to determine our sample size as shown in Equation 5.7. This equation is similar to Equation 4.1 used in survey sampling.

$$n = \frac{(z_{\alpha/2})^2 \sigma^2}{E^2} \qquad (5.7)$$

## Example
Suppose we suspect that 200 of a population of 1000 items are contaminated, and we want to determine the sample size to be examined with a maximum allowable error of 1% and confident level of 95%.

$$\sigma = p(1 - p) = (0.2)(0.8) \quad z_{\alpha/2} = 1.96 \text{ from the z tables} \quad E = 0.05$$

Substituting in Equation 5.7, $n = 245.86$. Hence 246 items are required to be sampled. If our suspicion of 20% contaminated is wrong, instead it was 30%. Replacing the value of $p$ with 0.3 and $q$ with 0.7, our new sample size will be 323. Similarly, if we are to increase the confident level to 99% , then with 20% contaminated the sample size required will be 426 items.

From this example we can appreciate the importance of having prior knowledge of our population to decide the sample size. When the information to be provided is the validation of a hypothesis, calculation of the sample size should be done using probabilities estimates rather than the mean estimates. The principal differences between the two methods are the number and form of errors. These formulae are mainly guides. In reality, there are other factors such as cost, degree of accuracy, presentation of data, time frame, and purpose that need to be considered in determining the size of a sample.

## 5.5 Interval Estimates

The central limit theorem introduced earlier in this chapter allow us to conclude that the sampling distribution of $\bar{x}$ and $p_{avg}$ can be approximated by a normal probability distribution whenever the sample size is large. As a result our sampling error can be equated to a probability. We can now present our estimate with a measure of precision. If we need to change the degree of precession, a larger sample size will be necessary. Using our normal distribution, we can say that there is a $(1 - \alpha)$ probability that the value of a sample mean will provide a sampling error of $z_{\alpha/2} \cdot \sigma_{\bar{x}}$ or less. Expressing this differently, we can say that there is a $1 - \alpha$ probability that the interval formed by $\bar{x} \pm z_{\alpha/2}\, \sigma_{\bar{x}}$ will contain the population mean $\mu$. In common statistical terminology, the intervals are called the confidence intervals, and the value $1 - \alpha$ is the confidence coefficient. We now combine the point estimate with the probability information about the sampling point estimate such as the mean to obtain an interval estimate.

### Example

A sample of 50 safety professionals in the country was used to decide the mean monthly salary of safety professionals. The sample provided a mean of \$3685 with a standard deviation of \$849. Expressing this results as a 90% confidence interval estimate of the population mean, and we use $\bar{x} \pm z_{\alpha/2}$ $\sigma_{\bar{x}}$. We know that $\bar{x} = 3685$, $\sigma = 849$, and $\alpha/2 = 0.05$. From our 'z' tables we can find $z_{\alpha/2}$ at 90%, which is equal to 1.645. So our interval estimate is \$3685 $\pm$ 1.645 $(849/\sqrt{50}) = \$3685 \pm 198$. We can now say with 90% degree of confidence that the monthly salary of some safety professional lies between \$3487 and \$3883. These calculations can be tedious. The good news is that computer software packages are available for computing confidence intervals for population estimates.

### Sample Variance

The sample variance can be calculated using the following Equation 5.8

$$s^2 = \frac{\sum (x_i - \bar{x})^2}{n-1} \tag{5.8}$$

Whenever a simple random sample of size $n$ is selected from a normal population, the quantity has a chi-square distribution with $n - 1$ degrees of freedom.

$$\frac{(n-1)s^2}{\sigma^2}$$

The interval estimate is given by Equation 5.9

$$\frac{(n-1)s^2}{\chi^2_{\alpha/2}} \leq \sigma^2 \leq \frac{(n-1)s^2}{\chi^2_{(1-\alpha/2)}} \tag{5.9}$$

where the $\chi^2$ values are based on a chi-square distribution of $n - 1$ degrees of freedom and $1 - \alpha$ is the confidence coefficient.

## Differences Between the Means of Two Populations

There are times when we will want to evaluate estimates from two different populations, for example, compare average accident rates of men versus women. To do this we apply the sampling distribution of $\bar{x}_1 - \bar{x}_2$ Equation 5.10.

$$\text{Expected value}: E(\bar{x}_1 - \bar{x}_2) = \mu_1 - \mu_2$$

$$\text{Standard deviation}: \sigma_{\bar{x}_1} - \sigma_{\bar{x}_2} = \sqrt{\frac{\sigma_1^2}{n_1} + \frac{\sigma_2^2}{n_2}} \tag{5.10}$$

where 1 and 2 are the different populations. If the sample sizes are large, then the sampling distribution of $\bar{x}_1 - \bar{x}_2$ can be approximated by a normal probability distribution. Two populations can be independent or matched. The statistical procedure is essentially the same in analyzing the difference of the means. Also, the statistical procedures for analyzing the differences between proportions of two different populations are similar to the procedure for analyzing the difference between means of two populations. Two independent estimates of $\sigma^2$ can be combined to generate a single estimate $\sigma_p$ by using the following Equation 5.11

$$\sigma_p = \frac{(n_1-1)*s_1^2 + (n_2-1)*s_2^2}{n_1+n_2-2} \tag{5.11}$$

# 6

## Testing and Inference

### 6.1 Introduction

Statistical inference is the process by which one draws conclusions regarding a population from the results observed in a sample. Hypothesis testing plays an important role in decision making. It is a more dynamic form of statistical inference than estimation. To introduce the notion, consider the following example: the trial of a new drug for the treatment of cancer at a specific body site. Experience and record analysis showed that patients with cancer at this site survive from first diagnosis to death, an average of 38.3 months, with a standard deviation of 43.3 months. It is hoped that the introduction of a new drug will prolong survival. An investigator tested this drug by administering it to a sample of 100 newly diagnosed cancer patients. These patients are followed until all die. The investigator determined that their mean survival duration is 46.9 months. What can we conclude regarding the drug's ability to prolong survival in this type of cancer? If this drug was adapted for the treatment of all patients with this cancer, will the results from the sample of 100 patients provide sufficient evidence that drug treatment on the average will prolong survival beyond the known mean of 38.3 months? In essence, can the investigator infer from a sample of 100 drug-treated cancer patients that the treatment will prolong survival duration. This inference that we wish to conclude can now be used to construct our hypothesis for testing. Hypothesis testing is used in many situations such as pure and applied research, validation of a claim, and decision-making situations.

### 6.2 Hypothesis Testing

We begin hypothesis testing by making an assumption about a population parameter. This assumption is called the null hypothesis and is denoted by $H_0$. We then define another hypothesis, called the alternative hypothesis, which is opposite of what is stated in the null hypothesis. The alternative hypothesis is denoted by $H_a$. We will discuss the tests for the mean ($\mu$) and proportion ($\pi$), since these are the two parameters most frequently used by

SHE professionals. From the last example, we assumed that treatment with the drug will prolong survival; therefore, we write $H_0: \mu > 38.3$ and $H_a: \mu = 38.3$. There are two possible outcomes: we can either accept or reject the null hypothesis ($H_0$). Frequently, we are uncertain whether or not the null hypothesis is true. The selected action is determined by applying a decision rule that specifies what to do for any computed level of the test parameter. Generally the null hypothesis is formulated based on the assumption that the statement is true. If we accept $H_0$ as true, and we are incorrect, then we made a Type II error. If we reject $H_0$ as true and we are incorrect, then we made a Type I error. Statisticians consider the probabilities of making these errors when setting decision rules. These probabilities are written as follow:

$$\alpha = Pr[\text{type I error}] = Pr[\text{reject } H_0 \mid H_0 \text{ true}]$$

$$\beta = Pr[\text{type II error}] = Pr[\text{accept } H_0 \mid H_0 \text{ false}]$$

Hypothesis testing is based upon the choice of error you prefer to accept. Normally Type I error is the preferred type. This does not mean that Type II error cannot be favored. The values of $\alpha$ and $\beta$ are related; therefore, the value of '$\alpha$' can be altered by changing the value of '$\beta$'; in so doing accepting that $H_0$ is less likely to be true. Both types of error probabilities can be reduced by increasing the sample size. Unfortunately, increasing the sample size may not always be possible since it is often dictated by circumstances. In testing a hypothesis we either accept or reject the null. Sometimes we are not able to accept or reject. The statement "do not reject $H_0$" is used in these cases to avoid accepting $H_0$ as true, thus eliminating the possibility of making a Type II error. This statement is equivalent to saying that our test results are inconclusive. In short, testing the validity of a claim, the null hypothesis is generally formulated on the assumption that the statement is true. In testing the validity of the hypothesis action is required if the null hypothesis is rejected or the alternative cannot be accepted.

## 6.3 Hypothesis Tests about a Population Mean — Large Sample

The methodology of hypothesis testing requires that we specify the maximum allowable probability of an error type. This is called the level of significance for the test, and is denoted by $\alpha$ for Type I error. We assumed the population and sample is normal, then using the normal distribution we calculate $z$, called the critical value. Recall from previous discussions in this book, the normal distribution has two tails. A test using one tail of the curve is called one-tailed; for example, values greater than mean plus two standard deviations. Let us itemize the hypothesis-testing procedure for the one-tailed tests about a population mean. We restrict ourselves here to the large-sample case ($n \geq 30$) where the central limit theorem permits us to assume a normal sampling distribution for the $x$. In a large-sample case, when $\sigma$ is unknown, we simply substitute the sample standard deviation $s$ for $\sigma$ in computing the test statistic. The general form of a lower-tailed test of a large-sample ($n > 30$) hypothesis tests about a population mean is formulated as:

$H_0 : \mu \geq \mu_0$ and

$H_a : \mu < \mu_0$ where the test statistic is

$$z = \frac{\bar{x} - \mu_0}{\sigma / \sqrt{n}} \tag{6.1}$$

Note that Equation 6.1 is similar to Equation 3.4. The difference between Equations 3.4 and 6.1 is Equation 3.4 is the sample $z$ value, while Equation 6.1 is the population $z$ value based on sample estimates. We reject the null hypothesis ($H_0$) if $z < -z_\alpha$ at $\alpha$ level of significance. If we are testing $H_0$: $\mu \leq \mu_0$ and $H_a$: $\mu > \mu_0$, then we reject $H_0$ if $z > z_\alpha$. In a large-sample case, the population standard deviation $\sigma$ is unknown, so we use sample standard deviation $s$ for computing the test statistic.

We can also use the probability of obtaining a sample result to test the hypothesis. Here the p-value table is used. The approach is similar. If the null hypothesis is true, the p-value is the probability of obtaining a sample result that is at least as unlikely as what is observed. The p-value can be used to decide for a hypothesis test by noting that if the p-value is less than the level of significance $\alpha$, the value of the test statistic must be in the rejection region. Similarly, if the p-value is greater than or equal to $\alpha$, the value of the test statistic is not in the rejection region.

## 6.4 Suggested Steps of Hypothesis Testing

1. Determine the null and alternative hypotheses that are appropriate for the problem.
2. Select the test statistic that will be used to decide whether to reject the null hypothesis.
3. Specify the level of significance for the test.

4. Develop the rejection rule that shows the values of the test statistic that will lead to the rejection of $H_0$.

5. Using the sample-computed statistic, compare it with the critical value(s) specified in the rejection rule to decide whether $H_0$ should be rejected.

So far we have tested one tail. The approach for testing a two-tailed hypothesis differs from a one-tailed test in that the rejection region is placed in both the lower and the upper tails of the sampling distribution. The general form for a large-sample hypothesis test about a population mean for a two-tailed test is written in the same manner as:

$$z = \frac{\bar{x} - \mu_0}{s / \sqrt{n}}$$

where $H_0$: $\mu = \mu_0$ and $H_a$: $\mu \neq \mu_0$. The rejection rule at a level of significance equal to $\alpha$ is, reject $H_0$ if $z_{\alpha/2} < z < -z_{\alpha/2}$. Similarly, the p-values can also be used to establish the rejection rule. A small p-value shows rejection of $H_0$, as with the one-tailed hypothesis tests. Again if $\sigma$ is unknown then we use $s$. The p-value is called the observed level of significance and it depends only on the sample outcome.

It is necessary to know whether the hypothesis test being investigated is one-tailed or two-tailed. Given the value of mean ($\bar{x}$) in a sample, the p-value for a two-tailed test will always be twice the area in the tail of the sampling distribution at the value of $\bar{x}$. The interval-estimation approach to hypothesis testing helps to highlight the role of the sample size. Larger sample sizes ($n$) lead to more narrow confidence intervals. Thus, for a given level of significance $\alpha$, a larger sample is less likely to lead to an interval containing $\mu_0$ when the null hypothesis is false. That is, the larger sample size will provide a higher probability of rejecting $H_0$ when $H_0$ is false.

## 6.5    Hypothesis Tests about A Population Mean — Small Sample

The methods of hypothesis testing that we discussed thus far required sample sizes of at least 30 elements. You recall the reason for this is the central limit theorem that enables us to approximate the sampling distribution of $\bar{x}$ with a normal probability distribution. When the sample size is small that is $n < 30$, the $t$-distribution is used to make inferences about the value of a population mean. In using the *t*-distribution for hypothesis tests about a population mean, the value of the test statistic is:

$$t = \frac{\bar{x} - \mu_0}{s / \sqrt{n}}$$

This test statistic has a *t*-distribution with $n - 1$ degrees of freedom. Note the similarities of the test statistics for large samples. The *t*-distribution is used instead of the *z*-tables.

## 6.6 Hypothesis Tests about a Population Proportion

Hypothesis tests about a population proportion are based on the difference between the sample proportion ($p$) and the hypothesized value ($p_0$). These are written as:

$$
\begin{array}{lll}
H_0 : p \geq p_0 & H_0 : p \leq p_0 & H_0 : p = p_0 \\
H_a : p < p_0 & H_a : p > p_0 & H_a : p \neq p_0
\end{array}
$$

where the first two columns are one-tailed tests and the third column is a two-tailed test. The methods used to conduct the tests are very similar to the procedures used for hypothesis tests about a population mean. The only difference is that we use the sample proportion $p$ and its standard deviation $\sigma_p$ in developing the test statistic. We begin by formulating null and alternative hypotheses about the value of the population proportion. Then, using the value of the sample proportion $p$ and its standard deviation $\sigma_p$, we compute a value for the test statistic $z$ as shown in Equation 6.2,

$$z = \frac{\bar{p} - p_o}{\sigma_{\bar{p}}} \tag{6.2}$$

where

$$\sigma_{\bar{p}} = \sqrt{\frac{p_o(1 - p_o)}{n}}$$

In small-sample cases, the sampling distribution of $p$ follows the binomial distribution and thus the normal approximation is not applicable. More advanced texts show how hypothesis tests are conducted for this situation. However, in practice small-sample tests are rarely conducted for a population proportion.

---

## 6.7  Hypothesis Tests about a Population Variance

In Chapter 5 Equation 5.9, we defined the sample variance as

$$s^2 = \frac{\sum (x_i - \bar{x})^2}{n-1}$$

and an interval estimate based on a chi-square distribution with $n - 1$ degrees of freedom. The value from the chi-square is used as the test statistics for the variance. For the one tailed-test $H_0: \chi^2 \leq \sigma_0^2$ and $H_a: \sigma_0^2 > \sigma_0^2$ and two-tailed, $H_0: \sigma^2 = \sigma_0^2$ and $H_a: \sigma^2 \neq \sigma_0^2$ where the subscript '0' is the hypothesized value. For one-tailed test we reject $H_0$ if $\chi^2 > \chi_\alpha^2$ and for the two-tailed if $\chi^2 < \chi_{(1-\alpha/2)}^2$ or if $\chi^2 > \chi_{\alpha/2}^2$.

---

## 6.8  Hypothesis Tests of Two Populations: Difference between Two Means

There are times when we will like to draw inferences from two different populations, for example we will like to compare the difference in the average accident rates of men and women in a company. These are two independent populations. The methodology again is divided into small and large cases. Recall in Chapter 5, Equation 5.10, the sampling distribution of $\bar{x}_1 - \bar{x}_2$ equations were given as:

$$\text{Expected value}: E(\bar{x}_1 - \bar{x}_2) = \mu_1 - \mu_2$$

$$\text{Standard deviation}: \sigma_{\bar{x}_1} - \sigma_{\bar{x}_2} = \sqrt{\frac{\sigma_1^2}{n_1} + \frac{\sigma_2^2}{n_2}}$$

where 1 and 2 are the different populations. If the sample sizes are large, then the sampling distribution of $\bar{x}_1 - \bar{x}_2$ can be approximated by a normal probability distribution. The hypotheses

$$H_0: \mu_1 - \mu_2 = 0 \qquad H_0: \mu_1 - \mu_2 \neq 0$$

2222

The test statistic z is given by Equation 6.3

$$z = \frac{(\bar{x}_1 - \bar{x}_2) - (\mu_1 - \mu_2)}{\sqrt{\sigma_1^2/n_1 + \sigma_2^2/n_2}} \tag{6.3}$$

Note if we are testing the null hypothesis, then $\mu_1 - \mu_2 = 0$, also for small samples we use the *t*-distribution.

## 6.9 Hypothesis Tests of Two Populations: Difference between Two Proportions

The approach is similar as the difference between two means. The hypotheses are written as:

$$H_0: p_1 - p_2 = 0 \qquad H_0: p_1 - p_2 \neq 0$$

and the test statistic is given by:

$$z = \frac{(\bar{p}_1 - \bar{p}_2) - (p_1 - p_2)}{\sigma_{\bar{p}_1 - \bar{p}_2}} \tag{6.4}$$

## 6.10 Hypothesis Tests of Two Populations — Comparing Variances

If we are comparing the variances of two independent samples then the ratio $s_1^2/s_2^2$ has a F-distribution if the populations' variances are equal. Thus, we can test for $F = s_1^2/s_2^2$. If $F > F_{\alpha/2}$ in the two-tailed test then $H_0$ is rejected and if $F > F_\alpha$ in the one-tailed test $H_0$ is rejected. The value of $F_{\alpha/2}$ is based on an F distribution with $n_1 - 1$ degrees of freedom for the numerator and $(n_2 - 1)$ degrees of freedom for the denominator.

## 6.11 Hypothesis Tests of More than Two Populations Means

A statistical technique called analysis of a variance (ANOVA) is used to test the hypothesis that the means of more than two populations are equal. To use this technique, the following conditions must be satisfied:

1.  For each population, the observations are normally distributed.
2.  The variances of the populations are equal.
3.  The observations are independent.

If the above conditions are satisfied, we can test for the equality of $L$ populations means, that is $H_0 : \mu_1 = \mu_2 \dots = \mu_L$ and the alternative hypothesis is that not all the means are equal. When the sample sizes are the same, the overall sample mean is the sum of the means divided by the number of populations. The between-samples estimate of the variance is called the mean square between and will be denoted by $EVB$ and is given by Equation 6.5

$$EVB = \frac{\sum_j^L n_j(\bar{x}_j - \bar{X})^2}{L-1} \tag{6.5}$$

where $\bar{x}$ is the overall sample mean, and $L-1$ represents the degree of freedom. The variation of the observations within each population is called the mean square within and will be denoted by $EVW$, and given by Equation 6.6

$$EVW = \frac{\sum_j^L (n_j - 1)s_j^2}{n_T - L} \tag{6.6}$$

where $n_T$ is the total of the samples from the populations. $EVW$ and $EVB$ provide *tw* independent and unbiased estimates of $\sigma^2$. Thus, if the null hypothesis is true then the sampling distribution of $EVB$ and $EVW$ is an $F$ distribution. If the means of the $L$ populations are not equal, the value of the ratio of $EVB$ to $EVW$ will be within the rejection area. So we reject the null hypothesis that the population means are not equal. If we are interested in determining where the differences among the means occur, there are several methods such as Fisher's Least Significant Difference, Tukey's Procedure, and Bonferroni Adjustment methods that can be used. Further discussion on this topic is beyond the scope of this book.

## 6.12 Hypothesis Testing and Decision Making

In testing hypotheses for research and validation of a claim, action is taken only when the null hypothesis $H_0$ is rejected, hence the alternative hypothesis $H_a$ is concluded to be true. In decision making it is necessary to take action whether the null hypothesis is accepted, rejected and not rejected. The hypothesis testing procedures presented thus far have limited applicability in a decision-making conclusion that $H_0$ is true. As previously mentioned, there are two types of errors. The probability of making a Type I error was controlled by establishing a level of significance for the test. The probability of making a Type II error was not controlled. In certain decision-making situations action may be required for both the conclusions "do not reject $H_0$" and "reject $H_0$." Type II error can be calculated using the rejection rule of Type I. The following is a step-by-step procedure that can be used in computing the probability of making a Type II error in hypothesis tests about a population mean.

1. Formulate the null and alternative hypotheses
2. Establish the rejection rule (use the $\alpha$ level of significance )
3. Solve for the value of the sample mean that identifies the rejection region for the test, and state the values of the sample mean that lead to the acceptance of $H_0$ (this defines the acceptance region for the test)
4. Using the sampling distribution of $\bar{x}$ for any value of $\mu$ from the alternative hypothesis, and the acceptance region, compute the probability of making a Type II error at the chosen value of $\mu$
5. Repeat the above step for other values of $\mu$ from the alternative hypothesis

Hypothesis testing is a statistical procedure that uses sample data to decide whether a statement about the value of a population parameter should be accepted or rejected. The hypotheses are two opposing statements, a null hypothesis $H_0$ and an alternative hypothesis $H_a$. The following guidelines are suggested for developing hypotheses based on two types of situations most frequently encountered in EHS.

1. **Validation of a claim** — In this situation the null hypothesis is true until proven incorrect. The claim made is chosen as the null hypothesis; the challenge to the claim is chosen as the alternative hypothesis. Action against the claim will be taken whenever the sample data contradict the null hypothesis.
2. **Decision making** — In this situation a choice between two courses of action must be made. One choice is associated with the null hypothesis and the other with the alternative hypothesis. It is suggested that the hypotheses be formulated such that the Type I error will be the more serious error. However, whenever an action must be taken based on the decision to accept $H_0$, the hypothesis-testing procedure should be extended to control the probability of making a Type II error.

The reliability of the test results can be improved by ensuring a minimum sample size given by Equation 6.7

$$n = \frac{(z_\alpha + z_\beta)^2 \sigma^2}{(\mu_0 - \mu_a)^2} \qquad (6.7)$$

where all of the terms are as previously defined. In a two-tailed hypothesis test, $z_\alpha$ is replaced with $z_{\alpha/2}$

## 6.13  Goodness of Fit Test

The chi-square distribution was introduced in Chapter 5 and illustrated how it could be used in estimation. The chi-square distribution could be used to compare the sample results with those expected when the null hypothesis is true. The most common statistical use of the chi-square distribution is the test for independence of two qualitative population variables. Two qualitative population variables $A$ and $B$ are independent if the proportion of the total population having any particular attribute of $A$ remains the same for the part of the population having a particular attribute of $B$, no matter which attributes are considered. The procedure for testing independence is based on the methodology used to test goodness of fit for a multi-nominal population. You first arrange the actual sample data in a contingency table involving one row $i$ for each attribute of the first variable and one column $j$ for each attribute of the second. The cell entries in this matrix, $f_{ij}$, are the actual frequencies for the respective attribute pairs. You compute the corresponding expected frequencies for the respective attribute pairs. Once you have found all $f_{ij}$ values, you may arrange them in a second contingency table. Matching the actual and expected frequencies from the respective cells, the test statistic for independence is given by Equation 6.8

$$\chi^2 = \sum_i \sum_j \frac{(f_{ij} - e_{ij})^2}{e_{ij}} \qquad (6.8)$$

where $e_{ij}$ = (Row $i$ total) (Column $j$ total)/sample size. All tests of independence are upper-tailed. Instead of using the normal distribution tables the chi-square table is used with the appropriate degree of confidence $\alpha$. The degrees of freedom is computed as the product of (number of rows $- 1$) times (Number of columns $- 1$). The worked example at the end of this chapter will outline a procedure to follow and illustrate the use. Note that a large chi-square value is an indication that the null hypothesis of independence should be rejected.

Thus, so far, you have encountered statistical testing procedures based on two sampling distributions: the normal and the Student-$t$. The chi-squared is a third distribution type which is useful for testing a variety of conditions. The most common statistical procedure involving the chi-square distribution is the test for independence of qualitative population variables. A large class of statistical procedures and applications uses the chi-square distribution. Like the Student-$t$, there is a different distribution for each value of $\chi^2$, depending on the number of degrees of freedom that applies for the particular test. One important chi-square application involves testing two qualitative variables for independence. Two qualitative population variables $A$ and $B$ are independent if the proportion of the total population having any particular attribute of $A$ remains the same for the part of the population having a particular attribute of $B$, no matter which attributes are considered.

If the computed value $\chi^2$ exceeds $\chi_\alpha^2$, you must reject the null hypothesis of independence; otherwise, you accept it. An analogous procedure involves testing for the equality of several proportions. You compute the test statistic in the same way, and you find the number of degrees of freedom and the critical value by an identical procedure. You can use the chi-square distribution to make inferences regarding the population variance and standard deviation. You use the expression in Equation 6.9 to construct a 100 $(1 - \alpha)$% confidence interval estimate of $\sigma^2$:

$$\frac{(n-1)s^2}{\chi_{\alpha/2}^2} \leq \sigma^2 \leq \frac{(n-1)s^2}{\chi_{1-\alpha/2}^2} \tag{6.9}$$

The critical values correspond to lower and upper tail areas of size $\alpha/2$, and the number of degrees of freedom is $df = n - 1$. If you take the square root of the limits, you obtain the confidence interval.

If you want to test the population variance, you compute the test statistic from the following expression in Equation 6.10. This test may be upper-tailed or lower-tailed.

$$\chi^2 = \frac{(n-1)s^2}{\sigma_0^2} \tag{6.10}$$

## 6.14 An Illustrative Example

The safety director of a chain of fast-food outlets showed that most accidents to customers occurred more often with older people than with younger people. To support this claim, a survey was conducted using the first 400 accidents that occurred. The results of the survey and a summary of the last five years are presented in Table 6.1 for comparison. We can use chi-squared tests to decide whether the safety director's claim could be correct or not.

**TABLE 6.1**

Age Distribution of Customers Involved
in an Accident in a Fast-Food Outlet

| Age Group | % of Customers Involved in an Accident | |
|---|---|---|
| | Summary | Survey |
| Under 20 | 5.0 | 6.5 |
| 20 to 30 | 10.0 | 15.7 |
| >30 to 40 | 15.0 | 16.0 |
| >40 to 50 | 20.0 | 20.5 |
| >50 to 60 | 30.0 | 23.5 |
| >60 to 70 | 15.0 | 14.0 |
| >70 | 5.0 | 3.7 |

First we calculate the chi-squared statistic as shown in Table 6.2.

**TABLE 6.2**

Calculations of Chi-Squared

| Age Group | Survey (S) | Expected (E) | $(S - E)^2$ | $(S - E)^2/E$ |
|---|---|---|---|---|
| <20 | 26 | 20 | 36 | 1.80 |
| >20 to 30 | 63 | 40 | 529 | 13.23 |
| >30 to 40 | 64 | 60 | 16 | 0.27 |
| >40 to 50 | 82 | 80 | 4 | 0.05 |
| >50 to 60 | 94 | 120 | 676 | 5.63 |
| >60 to 70 | 56 | 60 | 16 | 0.27 |
| >70 | 15 | 20 | 25 | 1.25 |
| Total | 400 | 400 | | 22.50 |

The chi-squared statistic equals 22.50 and is what we now use to test the significance of the claim. The degree of freedom is 6 since there were 7 age groups. Assume we test at the 95% level of confidence, then looking up the $\chi^2$ table the critical point value is 12.6. Hence our statistic exceeds the critical point value, so we reject the null hypothesis that is the claim.

## 6.15 Non-Parametric Methods

The statistical methods presented so far in this book are called parametric methods. This section deals with controlled experiments where the measurement scale is either nominal or ordinal and the populations are not normal or no assumption can be made. The Wilcoxon Signed-Ranked Test, the Mann-Whitney-Wilcoxon Test, the Kruskal-Wallis Test, and Spearman Rank Correlation are the non-parametric methods presented in this section.

## Comparing Two Samples

Traditionally, the sample from one population is called the control group. Evaluations with quantitative populations may be conducted using either two independent random samples or matched pairs involving partners from both groups. Two-sample comparisons may involve either estimates or tests. A variety of procedures are available for each type of inference. In comparing population means, you base your procedure on either the normal or the Student-$t$ distribution, depending on the sample size. You ordinarily use the normal approximation when comparing qualitative populations in terms of proportions. Employing non-parametric statistics when testing two quantitative populations may be advantageous. Two-sample investigations typically involve comparisons of two means or two proportions. You can make a comparison between two populations by estimating the difference of the means or, by conducting a hypothesis test to establish whether the mean or proportion for population $A$ is much greater or smaller than for population $B$.

For independently selected samples of large size, you establish the confidence interval estimate for the difference in means using Equation 6.11

$$\mu_A - \mu_B = \overline{X}_A - \overline{X}_B \pm \sqrt{\frac{S_A^2}{n_A} + \frac{S_B^2}{n_B}} \tag{6.11}$$

If you have ordered the observations into matched pairs, you compute the difference $d_i = X_{Ai} - X_{Bi}$ for each pair. You then find the mean difference $d$ and the standard deviation using equation 6.12.

$$\overline{\sigma} = \frac{\sum \sigma_i}{n} \qquad s_\sigma = \sqrt{\frac{\sum \sigma_i^2 - n\overline{\sigma}^2}{n-1}} \tag{6.12}$$

Finally, you use $d$ and $s_d$ to compute an interval estimate of the difference in means as given in Equation 6.13.

$$\mu_A - \mu_B = \bar{d} + z\frac{S_d}{\sqrt{n}} \qquad (6.13)$$

Testing procedures extend directly into two-sample case. The normal deviate z is the usual test statistic. For independently selected samples of larger size, you compute z from Equation 6.14

$$z = \frac{\overline{X}_A - \overline{X}_B}{\sqrt{\frac{s_A^2}{n_A} + \frac{s_B^2}{n_B}}} \qquad (6.14)$$

With matched pairs, the z value is obtained by using Equation 6.15

$$z = \frac{\bar{d}}{\frac{S_d}{\sqrt{n}}} \qquad (6.15)$$

When sample sizes are small, the Student-$t$ statistic replaces z. When constructing confidence interval estimates for the difference in means, use Equation 6.16 for independent samples.

$$\mu_A \mu_B = \overline{X}_A - \overline{X}_B + \tau_{\alpha/2}\sqrt{\frac{(n_A-1)s_A^2 + (n_B-1)s_B^2}{n_A + n_B - 2}} \cdot \sqrt{\frac{1}{n_A} + \frac{1}{n_B}} \qquad (6.16)$$

In testing hypotheses for the difference in means of small samples, use Equation 6.17 to compute the test statistic for independent samples.

$$t = \frac{\overline{X}_A - \overline{X}_B}{\sqrt{\frac{(n-1)s_A^2 + (n_B-1)s_B^2}{n_A + n_B - 2}} \cdot \sqrt{\frac{1}{n_A} + \frac{1}{n_B}}} \qquad (6.17)$$

For large samples the equation is the same except that $t$ is replaced with z.

In testing for the difference in proportions, you pool the samples to compute the combined sample proportion as shown in Equation 6.18

$$P_c + \frac{n_A P_A + n_B P_B}{n_A + n_B} \qquad (6.18)$$

The testing procedure employs the normal approximation, and you use $P_c$ in computing the test statistic z shown in Equation 6.19

$$z = \frac{P_A - P_B}{\sqrt{P_c(1 - P_c)\left(\dfrac{1}{n_A} + \dfrac{1}{n_B}\right)}} \qquad (6.19)$$

All these procedures are parametric in that they focus on population means or proportions. As an alternative, non-parametric statistics provides testing procedures that are free from some often unrealistic assumptions required by those methods.

For tests involving independent samples you may use the Wilcoxon Rank-Sum Test. You establish the test statistic by first ranking all observations, starting with the lowest value.

$$z = W - \frac{\dfrac{n_A(n_A + n_B + 1)}{2}}{\dfrac{\sqrt{n_A n_B (n_A + n_B + 1)}}{12}} \qquad (6.20)$$

In doing this you pool all the observations and assign ranks without regard to the sample group. You then compute the sum $W$ of the group $A$ ranks. The following test statistic applies. With matched pairs, you use the Wilcoxon Signed-Rank Test. This procedure is based on pair differences, which is ranked according to absolute value (ignoring the signs and any zero differences). Then you compute the rank sum $V$ for positive differences. You use the following test statistic:

$$z = \frac{V - \dfrac{n(n+1)}{4}}{\sqrt{\dfrac{n(n+1)(2n+1)}{24}}} \qquad (6.21)$$

*Section C*

# Information Generation

# 7

## Regression and Correlation

### 7.1 Introduction

Regression and correlation are two important mathematical associations used in statistics. These analyses are important tools in EHS applications of statistics especially when an unknown value needs to be predicted from the known values of other variables. Regression analysis is a statistical procedure that can be used to develop a mathematical equation showing how variables are related. The variable that is being predicted is called the dependent variable, while the variables used to predict the value is called the independent variable. Correlation analysis is concerned with the extent to which the variables are related, that is, the strength of the relationship. Regression and correlation analyses can only show how, or to what extent variables are associated with each other. Conclusions about cause and effect relationships must be based on the knowledge of the analyst.

### 7.2 Regression Equation

The least-square method is a procedure used to find the straight line that provides the best approximation for the relationship between the independent and dependent variables, say $x$ and $y$, respectively. These variables are plotted on a scatter diagram as shown in Figure 7.1 The regression equation can then be written as shown in Equation 7.1

$$Y(x) = a + bx \qquad (7.1)$$

where $a$ and $b$ are constants. The constant $b$ is the slope and calculated using Equation 7.2

$$b = \frac{n \sum XY - \left(\sum x\right)\left(\sum y\right)}{\left(n \sum x^2 - \sum x\right)^2} \qquad (7.2)$$

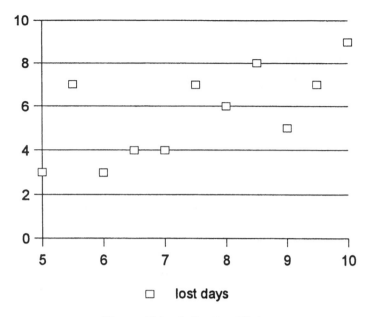

**Figure 7.1   A Scatter Plot.**

From the independent values, the dependent values (denoted by $y_r$) which best fit the line will be calculated. The differences between the actual and the best fit values will be discussed in the next section. The overall quality of the regression line is reflected by the standard error of the estimate, denoted by $S_{y \cdot x}$ and given by the following equation:

$$x_{Y \cdot X} = \sqrt{\frac{\sum Y^2 - a \sum Y - b \sum XY}{n - 2}} \qquad (7.3)$$

The value of $s_{y \cdot x}$ reflects the ability of the regression line to "explain" some variability in the values of $y$. A satisfactory regression line will yield a value for $s_{Y \cdot X}$ that is smaller than the sample standard deviation of $y$, denoted by $s_y$. The least squares method provides an estimated regression equation that minimizes the sum of squared deviations between the observed values of the dependent variable $y$ and the estimated values of the dependent variable $y_r$. This is the least squares criterion for choosing the equation that provides the best fit. If another criterion were used, such as minimizing the sum of the absolute deviations between $y$ and $y_r$, a different equation would be obtained. In practice, the least squares method is the most widely used.

## 7.3 Coefficient of Correlation

The coefficient of correlation denoted by $r$ provides a measure of the goodness of fit of the estimated regression equation to the data. The coefficient can be calculated using Equation 7.4.

$$r = \frac{n\sum xy - \sum x \sum y}{\sqrt{\left[n\sum x^2 - \left(\sum x\right)^2\right]\left[n\sum y^2 - \left(\sum y\right)^2\right]}} \qquad (7.4)$$

The value of $r$ is between –1 and +1. A zero value suggests no correlation. The square of the correlation coefficient is the coefficient of determination that is a measure of the proportion of the total variation in the value of $y$ that can be explained by the variations in the value of $x$. Using the regression line $y = a + bx$, the coefficient of determination $r^2$ is given by Equation 7.5

$$r^2 = \frac{a\sum y + b\sum xy - (1/n)\left(\sum y\right)^2}{\sum y^2 - (1/n)\left(\sum y\right)^2} \qquad (7.5)$$

The theory underlying regression and correlation analyses assumes that $x$ and $y$ are related by a true regression line and the data from the sample is a representation of the population. The coefficient of determination only establishes the relationship between the tested variables, for example, we examine number of workers ($x$) and the total lost time ($y$) for five months as shown in Table 7.1, and we calculated that there is a linear relationship given by $y = 28 + 2.6x$ and $r^2 = 0.98$, we can conclude that the lost time is related to the number of workers with a 98% confidence.

TABLE 7.1

Number of Workers and Total Lost-Time

| Month | Number of Workers | Total Lost Time (hrs) |
|---|---|---|
| Jan | 20 | 82 |
| Feb | 16 | 70 |
| March | 24 | 90 |
| April | 22 | 85 |
| May | 18 | 73 |

In developing the least squares estimated regression equation and computing the coefficient of determination, no probabilistic assumptions or statistical inferences have been made. Large values of $r$ simply imply that the least squares line provides a better fit to the data; that is, the observations are more closely grouped about the least squares line. Nevertheless, using only $r$, no conclusion can be made regrading whether the relationship between $x$ and $y$ is statistically significant. Such a conclusion must be based on considerations that involve the sample size and the properties of the appropriate sampling distributions of the least squares estimators. Typically values greater than 0.75 are considered useful in EHS.

## 7.4   Testing of Significance

An important idea that must be understood before we consider testing for significance in regression analysis involves the distinction between a deterministic analysis and a probabilistic analysis. In a deterministic analysis the relationship between the dependent variable $y$ and the independent variable $x$ is such that if we specify the value of the independent variable, the value of the dependent variable can be determined exactly. In the probabilistic analysis an exact is unable to be guaranteed. Regression analysis can be considered a probabilistic analysis. As a result we can rewrite our regression equation to include an error term $\varepsilon$. We assume that the error term is a random variable with mean or expected value equal to zero; the variance is the same for all values of $x$; the values of G are independent; and is a normally distributed random variable.

In Section 7.3 we saw how the coefficient of determination ($r^2$) could be used as a measure of the goodness of fit of the estimated regression line. Larger values of $r^2$ showed a better fit. However, the value of $r^2$ does not allow us to conclude whether a regression relationship is statistically significant. To draw conclusions concerning the statistical significance we must consider the sample size.

The standard error and deviation of the estimated regression line can be calculated using Equations 7.6 and 7.7, respectively.

$$se = \sqrt{\frac{\sum (y_i - y_e)^2}{n - 2}} \qquad (7.6)$$

where $y_e$ is the corresponding value of $y$ which would be estimated from the regression line.

$$S_b = se \sqrt{\dfrac{1}{\sum x^2 - \dfrac{\left(\sum x\right)^2}{n}}} \qquad (7.7)$$

## t Test and F Test

The null hypothesis for these tests is whether the estimated slope could have risen by statistical chance. Statistically, we use the ratio of the estimated value of $b$ to the standard deviation of the regression line as our t test value. The null hypothesis will be rejected if $t_{\alpha/2} < b_t < -t_{\alpha/2}$ where $b_t$ denotes the ratio. Another way of calculating our test statistic is by using Equation 7.8

$$r_t = \sqrt{\dfrac{r^2}{1-r^2}(n-2)} \qquad (7.8)$$

The t test and F test should yield the same conclusion of either to accept or reject the null hypothesis. However, for more than one independent variable only the F test should be used.

## 7.5 Estimation and Prediction

There are two types of interval estimates to consider. The first is an interval estimate of the mean valve of $y$ for a particular value of $x$. We refer to this type of interval estimate as a confidence interval estimate. It is given by $y_{avg} \pm t_{\alpha/2}s_y$ where the confidence coefficient is $1 - \alpha$ and t value has $n - 2$ degrees of freedom.

The second is the type of interval estimate used to predict an individual value of $y$ corresponding to a given value of $x$. We refer to this type of interval estimate as a prediction interval estimate. It is given by $y_{avg} \pm t_{\alpha/2}s_{ind}$ where $s_{ind}$ is given by Equation 7.9

$$s_{ind} = s \sqrt{\dfrac{1}{n} + \dfrac{(x-\bar{x})^2}{\sum x^2 - \left(\sum x\right)^2 / n}} \qquad (7.9)$$

for a particular value of $x$.

## 7.6   Residual Analysis

For each observation in a regression analysis there is a difference between the observed value of the dependent variable $y$ and the value predicted by the regression equation $y_r$. The analysis of residuals plays an important role in validating the assumptions made in regression analysis. In Section 7.5, we showed how hypothesis testing can be used to decide whether a regression relationship is statistically significant. Hypothesis tests concerning regression relationships are based on the assumptions made about the regression analysis. If the assumptions made are not satisfactory, the hypothesis tests are not valid, and the estimated regression equation should not be used. However, remember that regression analysis is an approximation of reality. There are two concerns in verifying that the assumptions are acceptable; namely, the four assumptions concerning the error term $\varepsilon$ and the form of the equation.

The residuals of $y - y_e$ are estimates of $\varepsilon$ that can help reveal patterns and thus determine whether the assumptions concerning G are satisfied. Three of the most common residual plots are:

1. Residuals against the independent variable $x$
2. Residuals against the predicted value of the dependent variable $y$
3. A standardized plot in which each residual is replaced by its z-score (i.e., the mean is subtracted and the result is divided by the standard error)

These plots can be used to identify observations that can be classified as influential or extraordinary. These influential or extraordinary observations may warrant careful examination. They may represent incorrect data, or simply unusual values that have occurred by chance. As a result we should first check to see if they are valid observations before isolating them. In regression analysis, it sometimes happens that one or more observations have a strong influence on the results obtained. Influential observations can be also identified from a scatter diagram when only one independent variable is present. Since influential observations can affect the estimated regression equation, it is important that they be examined carefully. A first check for the error is in collecting or recording the data. Influential observations may be valid and beneficial to our analysis. Once an observation has been identified as potentially influential because of a large residual, its impact on the estimated regression equation should be evaluated. More advanced texts discuss diagnostics for doing so. A simple procedure is to do the regression analysis with and without the observations. This approach can reveal the influence of the observations on the results.

## 7.7   Correlation Analysis

Sometimes the correlation between the two variables is of interest. The Pearson Product Moment correlation coefficient is used to measure the linear association. This is given by:

$$r_{xy} = \frac{s_{xy}}{s_x s_y} \tag{7.10}$$

where $s$ is the standard deviation of $x$ and $y$, $s_{xy}$ is the covariance and given by:

$$s_{xy} = \frac{\sum (x_i - \bar{x})(y_i - \bar{y})}{n-1} \tag{7.11}$$

This section noted that the sample correlation coefficient can also be tested for significance. Rejection of the null hypothesis does not permit us to conclude that the relationship between $x$ and $y$ is non-linear. It is valid to conclude that $x$ and $y$ are related and that a linear relationship explains a significant amount of the variability in $y$ over the range of $x$ values observed in the sample. Do not confuse statistical significance with practical significance. With very large sample sizes, it is possible to obtain statistically significant results for small values of $b$ that may not be practically significant. Regression and correlation analyses can only show how or to what extent the variables are associated with each other.

## 7.8   Multiple Regression

The analyses so far concentrated on two independent variables. There are activities when more than two independent variables must be analyzed. To do this, multiple regression analysis is used. It can be approached by applying the least squares method used in linear regression analysis. By using the least square approach, an estimated regression plane can be described in the form:

$$y = a + bx_1 + cx_2 + \ldots + nx_n \tag{7.12}$$

where $x_1$, $x2$, $x3$, … $x_n$ are the various factors affecting $y$. The function of $y$ will therefore be impossible to draw on a two-dimensional graph. The qualities of predictions reached in multiple regression analysis are usually better than in linear regression analysis. The same tests and coefficients calculated in the linear analysis can be done. These can be best done with the aid of a computer software program. Further treatment of multiple variables is beyond the scope of this book.

---

## 7.9  An Example

The safety director of company EIGHT Inc. believes that the funds allocated to the safety department and the yearly performance rating by management varies in direct proportion to the funds allocated. The rating maximum value is 40. The budgeted amount and rating value for the past six years are given in Table 7.2.

**TABLE 7.2**

Budgeted Amount and Rating Value for Company EIGHT, Inc.

| Year | Rating Value ($x$) | Allocated Funds ($y$) $,000 |
|------|--------------------|------------------------------|
| 1991 | 20 | 170 |
| 1992 | 19 | 170 |
| 1993 | 34 | 230 |
| 1994 | 25 | 200 |
| 1995 | 18 | 180 |
| 1996 | 22 | 190 |

   The average rating per year is 23 and funds allocated is $190,000. Calculating for $r$ using Equation 7.4

$$r = 0.96$$

Using Equation 7.2, the slope of the regression line $b = 3.693$, and

$$a = \text{average of } y - \text{slope} * \text{average } x$$

$$= 190 - 3.7 * 23$$

$$= 104.9$$

The regression line is $y = 104.9 + 3.7x$ as shown in Figure 7.2.

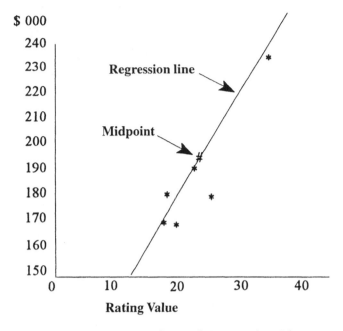

**Figure 7.2   Scatter Plot and Regression Line.**

The standard error of the estimate using Equation 7.6: $se = 6.98$ say 7. The standard deviation for the sampling distribution for the slope of the regression line using Equation 7.7, is 0.5276.

At 99% level of confidence with $(n - 2) = 4$ degrees of freedom, the $t$ statistic for $\alpha = 0.01$, our critical $t$ value = 4.604. So that the null hypothesis, that is the slope of the regression line $b$, arose by chance and that there is no significant differences between our estimated value of $b = 3.7$ and $b = 0$. Our $t$ statistic using the ratio of estimated slope to the standard deviation is $3.7/0.52 = 7.01$. Since our $t$ value is greater than our t critical value, we reject the null hypothesis and conclude that the slope of our regression line is significant at a 99% level of confidence. The standard deviation of $y$, for any given value of $x$, is found by using Equation 7.9, = 3.27.

# 8

## Time Series

### 8.1 Introduction

Time series is another type of quantitative forecasting. A time series analysis is based on historical data over a period of time concerning an activity or related to another time series. Usually it is used to forecast. In this chapter, we will discuss three time series methods of forecasting; namely smoothing, trend projection, and trend projection adjusted for seasonal influence. A graph of a time series is called a histogram, with the $x$ axis always representing the time, as shown in Figure 8.1. There are four features of a time series that are necessary to identify:

1. **Trend** — an underlying long term movement over time
2. **Seasonal variations or fluctuations** — short-term variations due to different circumstances that prevail and affect the results

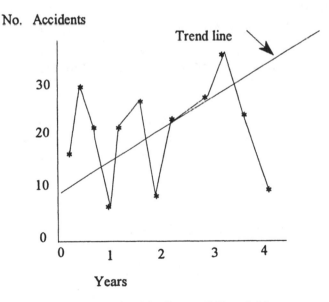

Figure 8.1  Time Series and Trend Line.

3. **Cycles or cyclical variations** — medium term changes in the results caused by factors that prevail for a while, then go away, then come back in a repetitive cycle

4. **Non-recurring, random variations** — unforseen circumstances resulting in unpredictable changes

The main problem we are concerned with when using time series analysis is how to identify the trend, seasonal variations, cyclical variations and random one-off causes in a set of data over a period.

## 8.2   Smoothing Method — Moving Average

There are three principal methods of separating a trend on a histogram. They are inspection discussed in Chapter 1, regression analysis by the least squares method covered in Chapter 7 and the moving averages. The moving average is an average taken at the end of each successive time period of the results of a fixed number of previous periods. Examples of lost time due to accidents for Company X are as shown in Table 8.1. The moving averages for 3 and 5 years period will be computed.

**TABLE 8.1**

Lost-Time Due to Accidents for Company X, 1990 to 1995

| Year | Lost Time (Days) | 3 Years Moving Average | 5 Years Moving Average |
|------|------------------|------------------------|------------------------|
| 1960 | 390 | | |
| 1962 | 380 | 410.00 | |
| 1963 | 460 | 430.00 | 430.00 |
| 1964 | 450 | 460.00 | 440.00 |
| 1965 | 470 | 450.30 | 464.00 |
| 1966 | 440 | 470.00 | |
| 1967 | 500 | | |

An example of calculation for a 3-year period:

$$\frac{(390 + 380 + 460)}{3} = 410$$

$$\frac{(380 + 460 + 450)}{3} = 430$$

An example of calculation for a 5-years period:

$$\frac{(390 + 380 + 460 + 470)}{5} = 430$$

Note that these values are related to the middle period as shown in Table 8.1 and the upward trend is more noticeable in the moving average series than the original time series. The choice of 3 or 5 years, or whatever period depends on the ability of the moving averages to isolate a trend.

The moving average's method can also be used to isolate seasonal variations. This can best be illustrated with an example. Suppose Company TJ operates from Wednesday to Sunday. The lost time due to injuries per day is carefully monitored for 4 weeks. Let us examine if there is a trend, and daily or weekly variations. The results of the monitoring are presented in Table 8.2.

**TABLE 8.2**

Lost-Time Injuries of TJ Company
Over 4-Week Period

| | Lost-Time (Minutes) | | | |
| | Week | | | |
| Day | 1 | 2 | 3 | 4 |
|---|---|---|---|---|
| Wednesday | 510 | 500 | 540 | 550 |
| Thursday | 360 | 380 | 390 | 410 |
| Friday | 570 | 058 | 580 | 600 |
| Saturday | 800 | 820 | 830 | 850 |
| Sunday | 850 | 840 | 870 | 900 |

Examining the data, a 5-days moving average is the obvious choice for us to be able to isolate weekly variations. The moving 5 days average and the difference between the actual value and the moving average are as presented in Table 8.3.

**TABLE 8.3**

Computation of Moving Averages and Differences

| Week | Day | Lost Time | Moving Total | Moving Average | Difference Between Moving Average and Actual |
|---|---|---|---|---|---|
| 1 | Wednesday | 510 | | | |
| | Thursday | 360 | | | |
| | Friday | 570 | 3090 | 618 | −48 |
| | Saturday | 800 | 3080 | 616 | +184 |
| | Sunday | 850 | 3100 | 620 | +230 |

**TABLE 8.3**    *(continued)*

Computation of Moving Averages and Differences

| Week | Day | Lost Time | Moving Total | Moving Average | Difference Between Moving Average and Actual |
|------|-----|-----------|--------------|----------------|-----------------------------------------------|
| 2 | Wednesday | 500 | 3110 | 622 | −122 |
|   | Thursday | 380 | 3130 | 626 | −246 |
|   | Friday | 580 | 3120 | 624 | −44 |
|   | Saturday | 820 | 3160 | 632 | +188 |
|   | Sunday | 840 | 3170 | 634 | +206 |
| 3 | Wednesday | 540 | 3170 | 634 | −94 |
|   | Thursday | 390 | 3180 | 636 | −246 |
|   | Friday | 580 | 3210 | 642 | −62 |
|   | Saturday | 830 | 3220 | 644 | +186 |
|   | Sunday | 870 | 3240 | 648 | +222 |
| 4 | Wednesday | 550 | 3260 | 652 | −102 |
|   | Thursday | 410 | 3280 | 656 | −246 |
|   | Friday | 600 | 3310 | 662 | −62 |
|   | Saturday | 850 | | | |
|   | Sunday | 900 | | | |

Now to examine the average daily variations we need to construct a table as shown in Table 8.4 that can clearly highlight the variations.

**TABLE 8.4**

Daily Variations of Lost Time Over 4-Week Period

| | Wed. | Thurs. | Fri. | Sat. | Sun. | Average |
|------|------|--------|------|------|------|---------|
| Week | | | | | | |
| 1 | | −48 | +184 | +230 | 122 | |
| 2 | −122 | −246 | −44 | +188 | +206 | −3.6 |
| 3 | −94 | −246 | −62 | +186 | +222 | 1.2 |
| 4 | −102 | −246 | −62 | | −136.7 | |
| Average | −106 | −246 | −54 | +186 | +219 | |

Applying the adjustments to our data and calculating the residues we can reduce the total variations. This trend, seasonal variation and predicted lost time as shown below are the elements that will be used in forecasting.

| Week No. | Day | Trend | Seasonal Variation | Predicted Lost Time | Actual Lost Time | Residual |
|---|---|---|---|---|---|---|
| 1 | Fri. | 618 | −54 | 564 | 570 | +6 |
| | Sat. | 616 | +186 | 802 | 800 | −2 |
| | Sun. | 620 | +220 | 840 | 850 | +10 |
| 2 | Wed. | 622 | −106 | 516 | 500 | −16 |
| | Thur. | 626 | −246 | 380 | 380 | 0 |
| | Fri. | 624 | −54 | 570 | 580 | +10 |
| | Sat. | 632 | +186 | 818 | 818 | +2 |
| | Sun. | 634 | +220 | 854 | 840 | −14 |
| 3 | Wed. | 634 | −106 | 528 | 540 | +12 |
| | Thur. | 636 | −246 | 390 | 390 | 0 |
| | Fri. | 642 | −54 | 588 | 580 | −8 |
| | Sat. | 644 | +186 | 830 | 830 | 0 |
| | Sun. | 648 | +220 | 868 | 870 | +2 |
| 4 | Wed. | 652 | −106 | 546 | 550 | +4 |
| | Thur. | 656 | −246 | 410 | 410 | 0 |
| | Fri. | 662 | −54 | 608 | 600 | −8 |

## 8.3   Forecasting

The difference between the actual data and the predicted data is the residual. The predicted data is the sum of trend data and the seasonal variation. The residual shows how much actual data were affected by external factors other than the trend and seasonal variations. Large residuals would reduce the reliability of the forecast. There are several mathematical techniques of forecasting. Techniques cannot eliminate uncertainty about the future, but they can help to ensure that considerations of currently known facts are included in the preparation of forecasts. Two methods will be illustrated with an example at the end of this section. The two methods are: predict values from the trend line and smooth the trend line by using the moving average for variations.

The first method can be illustrated using the above example. The trend line can be extrapolated to predict total lost time for the 6th week. This can be done either by summing the expected daily losses or using a weekly total plot. Either approach should result in values that are similar. The projected total for the 6th week in this example is equal to 732 lost minutes. More detail methods are available, fortunately such detailed calculations are not necessary in EHS for making sound financial recommendations or decisions. It is necessary that our trend line is as accurate as possible. This is the most critical property in using these simple techniques of forecasting.

## Illustrative Example

Using the information collected in 1990 to 1992 on the number of accidents for each quarter as shown in Table 8.5, we can predict the number of accidents for 1993:

**TABLE 8.5**

Number of Accidents by quarters for 1990 to 1992

| Year | Quarters | | | |
|------|-----|-----|-----|-----|
|      | 1st | 2nd | 3rd | 4th |
| 1990 | 18  | 30  | 20  | 6   |
| 1991 | 20  | 33  | 22  | 8   |
| 1992 | 22  | 35  | 25  | 10  |

The following is the suggested solution shown in Table 8.6.

**TABLE 8.6**

Solution to Worked Example on Accidents per Quarter

| Year | Quarter | Accidents | Moving 4-Quarter Total | Moving Average, 4-Quarter Total | Midpoints, Averages (Trend) | Difference Between Actual and Trend |
|------|---------|-----------|------------------------|---------------------------------|-----------------------------|-------------------------------------|
|      |         | (A)       | (B)                    | (B)/4                           | (C)                         | (A) to (C)                          |
| 1990 | 1st     | 18        |                        |                                 |                             |                                     |
|      | 2nd     | 30        |                        |                                 |                             |                                     |
|      | 3rd     | 20        | 74                     | 18.5                            | 18.75                       | +1.25                               |
|      | 4th     | 6         | 76                     | 19                              | 19.375                      | +13.375                             |
| 1991 | 1st     | 20        | 79                     | 19.75                           | 20                          | 0                                   |
|      | 2nd     | 33        | 81                     | 20.25                           | 20.5                        | +12.5                               |
|      | 3rd     | 22        | 83                     | 20.75                           | 21                          | +1                                  |
|      | 4th     | 8         | 85                     | 21.25                           | 21.5                        | −13.5                               |
| 1992 | 1st     | 22        | 87                     | 21.75                           | 22.125                      | −0.125                              |
|      | 2nd     | 35        | 90                     | 22.5                            | 22.75                       | +12.25                              |
|      | 3rd     | 25        | 92                     | 23                              |                             |                                     |
|      | 4th     | 10        |                        |                                 |                             |                                     |

Seasonal variations from the trend are shown in Table 8.7

**TABLE 8.7**

Seasonal Variations

| Year 1 | 1st Quarter | 2nd Quarter | 3rd Quarter | 4th Quarter | Total |
|--------|-------------|-------------|-------------|-------------|--------|
|        | —           | —           | +1.25       | −13.375     | −12.125 |
| Year 2 | 0           | +12.5       | +1          | −13.5       | 0 |
| Year 3 | −0.125      | +12.25      |             |             | +12.125 |
| Average | −0.625     | +12.375     | +1.125      | +13.4375    | 0 |

Since the number of accidents is integers, we should round up our numbers to one decimal place, therefore the seasonal variations are:

1st quarter  −0.1
2nd quarter  +12.4
3rd quarter  +1.1
4th quarter  −13.4.

The trend line suggests an increase of about 0.6 per quarter, calculated from the average quarterly increase in trend line values between the 3rd quarter of year 1 (18.75) and the 2nd year (22.75), that is (22.75 − 18.75)/7 = 4/7 = 0.57. Taking 0.6 as the quarterly increase in the trend, the forecast of accidents for 1993, before and after seasonal adjustments are as shown in Table 8.8.

**TABLE 8.8**

Predicted Values for 1993

| Year | Quarter | Trend Line | Average Seasonal Variation | Forecast Accidents |
|------|---------|------------|----------------------------|--------------------|
| 1992 | 2 | 22.8 | | |
|      | 3 | 23.4 | | |
|      | 4 | 24.0 | | |
| 1993 | 1 | 24.6 | −0.1 | 24.5 |
|      | 2 | 25.2 | 12.4 | 37.6 |
|      | 3 | 25.8 | 1.1 | 26.9 |
|      | 4 | 26.4 | −13.4 | 13.0 |

## 8.4   Exponential Smoothing

Exponential smoothing is a method for short-term forecasting. It is given by the formula:

new forecast = old forecast = $\alpha$ (most recent observation – old forecast)

where $\alpha$ is the smoothing factor or the smoothing constant. The closer $\alpha$ is to 1, the greater will be the emphasis on recent data in preparing a new forecast, whereas the closer to $\alpha$ is to 0 the greater will be the emphasis on the old forecast. In exponential smoothing, a new forecast takes a balanced view of the most recent actual data and the previous forecast. The importance of the previous forecast diminishes as a new forecast is made; therefore, the exponential label to the method.

# 9

## Linear Programming

### 9.1 Introduction

One of the most important decisions faced by EHS managers is the allocation of limited resources of manpower, materials, machinery, and money. In most businesses, allocation of resources is based on cost benefit or best advantage. Linear programming is a technique used to solve and determine the best combination within the restrictions or constraints that exist. There are two components in developing this model: an objective function and constraints. The major assumption used in linear programming is that the variables are additive and divisible, which are direct proportions. The objective function is a quantified statement of what are the best results or best advantage that is being aimed for as the objective of the resource allocation decision. Constraints are the limits of the conditions and variables, for example, maximum number of operating hours per day.

### 9.2 Formulating a Linear Programming Problem

Consider Joe Company that undertook a contract to supply a customer with at least 260 units in total of two products, X and Y, during the next month. At least 50% of the total output must be units of X. The products are each made by 2 grades of labor, as follows:

| Product | X | Y |
|---|---|---|
| Grade A labor | 4 hours | 6 hours |
| Grade B labor | 4 hours | 2 hours |

Although additional labor can be made available at short notice, the company wishes to use 1200 hours of Grade A labor and 800 hours of Grade B labor that has already been assigned to working on the contract next month. The total variable cost per unit is $120 for X and $100 for Y. Joe Company wishes to minimize expenditure on the contract next month. How much of X and Y should be supplied to meet the terms of the contract?

First we need to identify the variables, the objective function, and the constraints. Let the number of units of X supplied be $x$, and the number of units of Y supplied be $y$. Then the objective function is to minimize the cost to produce $x$ and $y$ that is, to minimize $120x + 100y$ subjected to the constraints. The constraints are:

$$x + y > 260 \qquad \text{(supply total)}$$

$$x > 0.5(x + y) \qquad \text{(proportion of } x \text{ in total)}$$

$$4x + 6y > 1200 \qquad \text{(Grade A labor)}$$

$$4x + 2y > 80 \qquad \text{(Grade B labor)}$$

$$x, y > 0$$

Note we can simply the constraint $x > 0.5 (x + y)$ to $x > y$.

## 9.3   Solving the Problem

Note that the constraints are expressed in inequalities, and not equations. Thus for a non mathematical person the graphical approach to linear programming is recommended. Three steps are involved in the graphical process:

1. Drawing the constraints in the problem graph
2. Identifying the feasibility area
3. Identifying the combination within the feasibility area.

Continuing to solve our example above, Figure 9.1 shows the constraints and identifies the feasible region.

The boundary line is the limiting function between the variables, for example in this equation it is $x = y$ since $x \geq y$. Our objective is to minimize cost, given by $120x + 100y$. The cost is minimized somewhere in the feasible region. This occurs at a turning point in the region. In this example it occurs where the constraint line $4x + 2y = 800$ crosses the line $x + y = 260$. We can read these values from the graph, or solve it mathematically. Either way should result in the same answer.

Solving mathematically:

$$x + y = 260 \qquad \ldots(1)$$

$$4x + 2y = 800 \qquad \ldots(2)$$

$$x = 140 \text{ and } y = 120$$

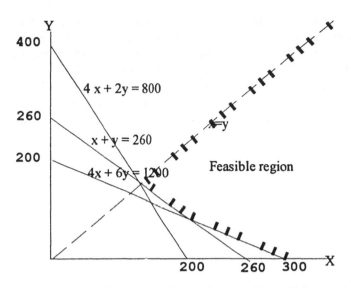

**Figure 9.1   Constraints, Boundary Line, and Feasible Region.**

Costs will be minimized by supplying:

|                | Unit Cost ($) | Total Cost ($) |
| -------------- | ------------- | -------------- |
| 140 units of X | 120          | 16,800         |
| 120 units of Y | 100          | 12,000         |
|                |               | $28,800        |

The proportion of units of X in the total would exceed 50%, and demand for Grade A labor would exceed the 1200 hours minimum.

We can now present further information to our problem by examining the effects of changes to variables to our solution. This is called sensitivity analysis. For example, suppose we increase the number of unit $x$ to 141 and decrease y to 119, what will be the effect? Clearly the cost will increase by $20, and we will not use a minimum of 1200 hours of grade A labor.

There are several limitations to this technique:

1. The development of the constraints in definite terms
2. The assumption on linearity
3. The model is static that is changes in the variables are not considered
4. Errors that may be caused by mathematical calculations, such as rounding up.

In reality, most problems involved more than two variables. These are solved by the simplex method. The simplex method solves for variables by using matrix mathematics. These are best solved by using a computer; however, you will be responsible for deciding the objective function and constraints. In conclusion, linear programming is a useful and powerful tool for optimizing resources if the assumptions are stated and understood: it can help to provide better budgeting decisions when resources are scarce.

# 10

## Network and Gnatt Chart

### 10.1 Introduction

Network analysis is a technique used to control and plan large projects. The primary aim of network analysis is to monitor the progress of a project and plan for the completion of a project in a given time. In so doing, it pinpoints the part of the project that is crucial and critical for completion on time. It can also be used in the allocation of resources such as labor and equipment. Thus, the optimizations of resources and time are the functions of network analysis to be discussed in this chapter.

A project can be divided into different tasks. These tasks are single activities that must be accomplished; for example inconstructing a new building, one task is to identify the site. After identifying the tasks, the order or sequence must be decided on, followed by the time required to complete each task. Network analysis is operated in various forms under a number of titles such as critical path analysis (CPA), projects evaluation and review technique (PERT) and critical path method (CPM). The technique is simple and graphical.

### 10.2 Network Analysis Techniques

There are five aspects to the network analysis covered in this chapter:

1. Drawing a network
2. Analyzing a network
3. Calculating float times
4. Determining uncertain activity times
5. Allocation of resources

#### Drawing the Network

A task or activity is represented by an arrowed line running between one event and another event. An event is the starting or ending point of a task. The normal flow of tasks in the diagram is from left to right as shown in Figure 10.1 The task or activity will be denoted by letters of the alphabet and the events by numbers.

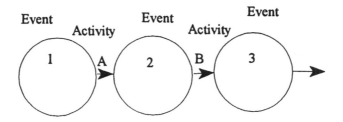

**Figure 10.1   Network Illustration.**

Sequencing the activities is our next step, for example, some activities cannot begin until others are completed. These are the preceding activities of the activity. Let us review an example to highlight the key points in drawing the network. A project consists of 12 activities in sequence as shown in Table 10.1.

**TABLE 10.1**

List of Activities and Sequence

| Activity | Preceding Activity |
| --- | --- |
| A | — |
| B, C, D | A |
| E, F | B |
| G | E |
| H | F |
| J | C |
| K | D |
| I | G, H |
| L | I, J, K |

To draw the network note the following:

1. Start with earliest activity on the left
2. Two activities should not be drawn between two same events
3. No fixed order in labeling activities and events, except the first event should be numbered with the least number and the last event with the largest number

The resulting network is shown in Figure 10.2.

Whenever it is not possible to start and end two activities with two different events, a dummy activity is introduced and represented by a dotted line. Also, a dummy line can be used to preserve the logic of the network. The times required to complete each activity given in Table 10.2 can now be inserted below or above the activity lines, as shown in Figure 10.3.

**TABLE 10.2**

List of Activities and Duration

| Activity | Duration (Days) | Activity | Duration (Days) |
|----------|-----------------|----------|-----------------|
| A | 5 | G | 4 |
| B | 4 | H | 3 |
| C | 4 | I | 2 |
| D | 3 | J | 1 |
| E | 5 | K | 4 |
| F | 5 | L | 3 |

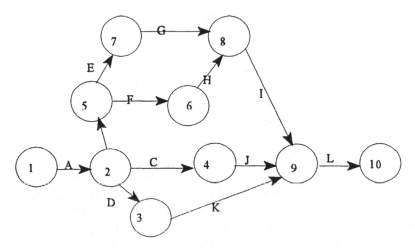

**Figure 10.2   Network of Worked Example.**

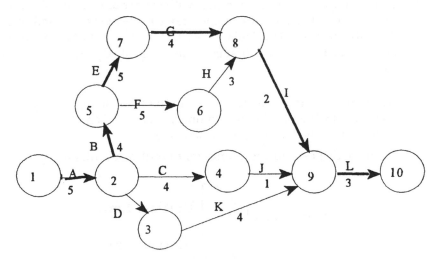

**Figure 10.3   Presentation of Event Node.**

## Analyzing the Network

Now we are ready to analyze the network to decide the critical path and minimum length of the project. A network can be analyzed into a number of different paths or routes. A path is a sequence of activities that can take you from start to end of the network. The duration of the project will be fixed by the time taken to complete the largest path through the network. This path is called the critical path and activities on it are called the critical activities. In the example, the paths and time to complete them are as follow:

| | |
|---|---|
| A to C to J to L | 13 days |
| A to B to E to G to I to L | 23 days |
| A to B to F to H to I to L | 22 days |
| A to D to K to L | 15 days |

Therefore, paths A to B to E to G to I to L is the critical path, and indicated by the thick lines on Figure 10.3 Activities on the critical path must start and end on time, otherwise the total project time will be extended.

## Calculating Float Times

The earliest and latest event times can now be inserted on the network. The earliest event time is the earliest time that any subsequent activities can start. The latest event time is the latest time that any preceding activity must be completed if the project as a whole is to be completed in the minimum time. This can be shown on the event nodes by dividing the circle into three sections as illustrated in Figure 10.4.

To calculate the earliest times, we start with event number one. The time at this event is zero. In a logical sequence move to the next event and so on throughout the diagram. To calculate the latest times, we start with the last event and proceed toward the first event in a logical sequences. Note the following:

a.  When more than one activity ends in the same node, the largest value controls when the earliest time is calculated and the lowest value when the latest time is calculated.

b.  Both the earliest and latest times for the critical events are equal.

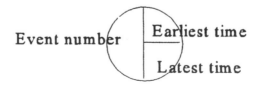

**Figure 10.4    Presentation of Event Node.**

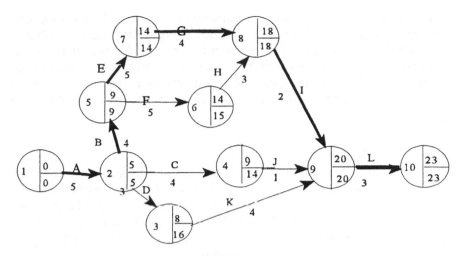

**Figure 10.5   Earliest and Latest Times of Worked Example.**

The earliest and latest times of the example are shown in Figure 10.5. The difference between the latest and earliest starting times is called a float. There are no floats in critical activities. Floats can be categorized into three different types: total, free, and independent. Total float is equal to the latest finishing time at the completion event minus the sum of the earliest starting time and the duration. This is the time by which its duration could be extended before it becomes critical.

Free float is equal to its earliest finishing time minus the sum of the earliest starting time plus the duration. This is the amount by which its duration can be increased without affecting either the total project time or the time available for subsequent activities. Independent float is equal to the earliest finishing time minus the sum of the latest starting time and the duration. This is the amount by which an activity could be extended without affecting the total project time, the time available for subsequent activities or the time available for preceding activities. The floats in this example are shown in Table 10.3

**TABLE 10.3**

Float Times of Example

| Activity | Float | | |
|----------|-------|------|-------------|
|          | Total | Free | Independent |
| C | 10 | 0 | 0 |
| D | 8 | 0 | 0 |
| K | 8 | 8 | 0 |
| J | 10 | 10 | 0 |
| F | 3 | 0 | 0 |
| H | 1 | 1 | 0 |

We can use these float times to minimize wastage or reallocate resources. For example, activity D can start as much as eight days after the 5th day without affecting the completion day of the project. However if activity D starts eight days later, then activity K is now a critical activity. A project can only be shortened by adjusting one activity on the critical path. The time required to perform non-critical activities is irrelevant as affecting the project time.

## Uncertain Activity Times

Activity times are not always easily determined. A probabilistic approach may be necessary; for example, the probability of completing activity A in 5 days is 0.6, 6 days is 0.2 and 7 days is 0.2. We can now use the expected time that is, (5 * 0.6) + (6 * 0.2) + (7 * 0.2) = 5.6 days to draw our network. We could also find the minimum and maximum times by using the appropriate days, and ignoring the probabilities.

PERT introduces a form of time-uncertainty analysis by assigning three time estimates to each activity, namely optimistic ($o$), most likely ($m$), and pessimistic ($p$). These estimates are converted into a mean time and a standard deviation. The mean time is given by:

$$\mu = \frac{o + 4m + p}{6} \tag{10.1}$$

and the standard deviation is given by:

$$\sigma = \frac{p - o}{6} \tag{10.2}$$

Once the mean and the standard deviation of the expected time have been calculated for each activity, estimating the critical path and the standard deviation of the expected total project time would be possible. Frequently the results of PERT are similar to other network analyses.

## 10.3 Allocation of Resources

Gnatt charts can be used to estimate the amount of resources required for a project. Consider our example with the number of workers required to achieve the activity in the given duration shown in Table 10.4.

Using Figure 10.5, the Gnatt chart will be drawn as shown in Figure 10.6 with the critical path as the basis.

**TABLE 10.4**

Activities, Duration, and Number
of Workers

| Activity | Duration (Days) | Number of Workers |
|:---:|:---:|:---:|
| A | 5 | 10 |
| B | 4 | 8 |
| C | 4 | 12 |
| D | 3 | 15 |
| E | 5 | 10 |
| F | 5 | 12 |
| G | 4 | 8 |
| H | 3 | 3 |
| I | 2 | 10 |
| J | 1 | 15 |
| K | 4 | 5 |
| L | 3 | 12 |

It can be seen that if all activities start at their earliest times, as many as 37 workers would be required on the 9th day, whereas on other days there would be excess workers. Excess workers can be reduced or removed by using up the float time on non-critical activities. For example, we can delay starting activities I and K, this will reduce the maximum number of workers required at anyone time from 37 to 22. Also, other activities can be planned for the idle labor. Similar manipulations can be done with other resources. Remember: the total project time is not affected unless a critical activity is adjusted. Network analysis is a very useful tool that can aid management in making decisions.

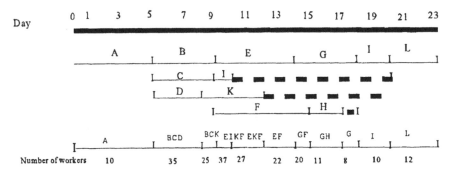

Figure 10.6   Gnatt Chart of Worked Example.

# 11

## Decision Analysis

### 11.1 Introduction

Decision analysis or operations research consists of a large range of different techniques each suitable for different applications. It can be used to decide optimal strategies when a decision maker is faced with several decision alternatives and an uncertain or risk-filled pattern of future events. What is most important is the approach rather than the actual technique used. Four major points of decision analysis should be remembered:

1. The approach is to develop a scientific model of the system under investigation to forecast or aid in making decisions or strategies.
2. It is essentially an aid to decision making.
3. It is based on quantifiable variables and logic rather than qualitative variables.
4. There are limitations and boundaries.

Operations research projects are normally a team effort involving accountants, technical experts and operational workers. This chapter introduces decision analysis by considering the stages in an operations research project. It then presents a summary of the success application of regression modeling by a company. The identification and information that can lead to the identification of the company were changed to protect the rights of the company. Model building in regression analysis is the process of developing a regression model that best describes the relationship among the independent and dependent variables. An option often overlooked is the decision to seek additional information through experimentation or observations. Also, discussed are the major pitfalls of experimentation and observations. The tools used in formulating decision models were discussed in previous chapters.

## 11.2 Stages in a Decision-Making Process

The main stages in the decision-making process of a research project are as follows:

1. Definition of the problems and objectives.
2. Analysis of the system to identify variables and states of nature.
3. Construction of the model.
4. Manipulation and testing the model.
5. Implementation.

## 11.3 Definition of the Problem

Clearly defining the objectives of the study or research in measurable terms that can form the basis for decision making is the first step in decision analysis. In formulating the problem statement and objectives, one must be careful not to create a problem that does not exist.

## 11.4 Analysis of the System

In analyzing the system, one may have an idea of the variety of possible future events. However, it must be remembered that you are unsure which particular event will occur. Thus, the second step in a decision-analysis approach is to identify the variables and future events that might occur. The future events, which are not under your control, are called the states of nature. It is assumed that the list of possible states of nature includes everything that can happen, and that the individual states of nature do not overlap; that is, the states of nature are defined so that one and only one listed state of nature will occur. In decision analysis terminology, the outcome resulting from making a certain decision and the occurrence of a particular state of nature is called the payoff. The variables are inputs and outputs that may be known with certainty or uncertainty. Again, these variables may be controllable or constants. Most real-life situations are complex and they are influenced by many variables. Fortunately, a few variables dominate the system. These are the variables and their interrelationship that you need to identify and build your model on. There are no fixed rules for deciding identifying variables; however, accuracy requires that as much depth and width of the variables be collected.

### Experimental and Observational Pitfalls

Besides the points discussed in Chapter 4 on sampling methods, the uniformity in the study subjects property of both experimental and observational data is important. Variations sometimes obscure systematic influences among groups, also making it difficult to isolate reasons for observed differences. In an experimental setting, individuals chosen to be as similar as possible are randomly assigned to comparison groups, diminishing the influence of extraneous variation. Rarely, observational data are controlled to the same extent. Observational studies rely on statistical techniques to deal with differences that result from lack of randomization and to reduce of extraneous variation. Techniques such as matching, stratifying, and statistical modeling are typical analytic approaches to observational data.

## 11.5  Constructing the Model

A central part of decision analysis is constructing a model of the real-world situation under study. A model is frequently used because it is impossible to experiment with the physical, real-world system itself. Before a model can be constructed, a study of the "real world" should be done. One approach is to develop a simple model and build on it. Models, can usually be divided into many types. The models we are interested in are mathematical: resource allocation, inventory, and competition. Note that experts in problem-solving approaches agree that the first step in solving a complex problem is to decompose it into a series of smaller subproblems. Decision trees provide a useful means of showing how the problem can be decomposed, and showing the sequential nature of the decision process. It provides a graphical representation of the decision-making process. People often view the same problem from different perspectives. Thus, there is no one correct way to develop a decision tree for a problem.

Occasionally we need to consider models that do not require knowledge of the probabilities of the states of nature. These approaches are appropriate in situations where there is very little confidence in the probabilities of the various states of nature, or where considering best is desirable and worst-case analyses that are independent of the probabilities of the state of nature. Different decision models sometimes lead to different decision recommendations. As a result, an understanding of the states of nature is important. This will enable you to relate to the complexities of the problem and select an appropriate objective, such as to optimize, to conserve or to reduce. Also, having a clear understanding of the problem will guide our outlook of the decision, such as optimistic, conservative, or min–max. The optimistic outlook will evaluate each decision alternative about the best outcomes that can occur. The conservative outlook will evaluate each decision alternative as for the worst outcomes that can occur. The min–max regret outlook is neither purely optimistic nor purely conservative. It identifies lost opportunities. Note again, that these different outlooks can lead to different recommendations.

## 11.6 Manipulation and Testing of Model

It is essential that the model and its solution are adequately tested before implementation. Controls should be established and sensitivity analysis performed. Sensitivity analysis is the study of the effect of such changes on the states of nature. The benefit of doing sensitivity analysis is that it can provide a better perspective on original judgement regarding the state-of-nature probabilities.

In many decision-making situations, obtaining probability estimates for each of the possible states of nature is possible. When such probabilities are available, the expected value approach can be used to identify the best decision alternative. The expected value approach evaluates each decision alternative about its expected value. The decision alternative recommended is the one that provides the best expected value.

Although many decision problems are complex, it is advisable to first draw a decision tree consisting of decision and state-of-nature nodes and branches that describe the sequential nature of the problem. If the expected value approach is to be used, the next step is to decide the probabilities for each of the state-of-nature branches and compute the expected value at each state-of-nature node. The decision branch leading to the state-of-nature node with the best expected value is then selected. The decision alternative associated with this branch is the recommended decision.

Frequently, decision analyses are conducted with sample information and preliminary or prior probability estimates for the states of nature. These are initially the best probability values available. However, to decide, additional information should be obtained about the states of nature. This new information can be used to revise or update the prior probabilities so that the final decision is based on more accurate probability estimates for the states of nature. The obtaining of additional information is most often accomplished through experiments designed to provide sample information or more current data about the states of nature.

## 11.7 Implementation

The implementation of the solution is obviously important. It is useful for the solutions to be implemented by the decision analysis team. Also, preparing a set of operating instructions for the new system is advisable.

## 11.8 Application of Regression Modeling

The challenges in regression modeling are in finding a proper form of the relationship and selecting variables. Temple Company began 80 years ago in the midwest, manufacturing one chemical for use in agriculture. Today, Temple is one of the nation's largest chemical companies, producing more than 1000 products ranging from industrial chemical products to food supplements. It is now an international company with manufacturing facilities, laboratories, technical centers, and marketing operations in 58 countries.

The Food Chemicals Department of the company manufactures and markets a food supplement for use by the poultry industry. The poultry industry is noted for its high volumes and small profit margins. As a result, they have specific nutritional requirements for poultry. Optimal feed compositions result in more rapid growth and higher final body weight for a given feed supply. Temple has worked closely with poultry growers to optimize poultry feed products. This working relationship has resulted in the low cost of poultry compared to other meats.

In studies conducted by Temple, regression analysis has been used to relate body weight after a fixed period of growth say $y$, to the amount of supplement added to the feed, say $x$. Initially a simple linear model was developed. The simple linear model proved statistically significant and explained 78% of the variability in the dependent variable. The straight-line relationship provided by the estimated regression equation suggests that increases in the level of the supplement will result in increases in the weight. Further research conducted by Temple has shown that at some point further increases in the level of the supplement led to a leveling off in body weight. As a matter of interest, it was found that when the amount of the supplement increases beyond nutritional requirements, the body weight may decline. As a result, Temple concluded that the quantity of the supplement that provides peak growth performance cannot be found using a straight-line relationship. Residual analysis suggested that a curvilinear relationship might provide a better model. As a result further experimentations were conducted and additional data were collected.

In the further studies, Temple used a nonlinear regression model to account for the curvilinear pattern in the data. It was found that the optimal body weight occurs when two variables are satisfied. This permitted Temple Company to use the information to optimize the benefits and minimize wastage of it's food supplement.

## 11.9 Conclusion

Models used in most statistical analyses are carefully constructed to reflect the observed data, and provide a mathematically convenient way to deal with complex issues without detailed knowledge of underlying mechanisms. The relationships surrounding the occurrence of occupational illness, for example, are not governed by mechanisms with simple mathematical expressions. However, simple mathematical expressions are often useful for summarizing, analyzing, and understanding these relationships. The importance of selecting an appropriate model cannot be overemphasized.

Computer software packages for decision analysis are readily available. Some allow the user to develop a graph of the decision tree on the screen and do all the decision-analysis calculations. In this chapter, decision analysis was outlined. The goal of decision analysis is to identify the best decision alternative given an uncertain or risk-filled pattern of future events such as accidents. The five stages in decision analyses were briefly discussed. Three outlooks to decision making were mentioned. While we mostly think only about situations with a finite number of states of nature, there will be situations where the states of nature are numerous. It will then be impossible or impractical to treat the states of nature as a discrete random variable consisting of a finite number of values. In such cases, more advanced techniques are required. These are beyond the scope of this book. Before attempting to use a decision model, the reader is advised to do additional reading on the subject. See "Selected Bibliography" at the end of this volume.

# 12

## Project Examples

### 12.1 Introduction

The following six examples are presented to highlight the use of quantitative methods in EHS projects. Each example will highlight subjects discussed in the previous chapters. While there are many software packages availablefor doing the calculations required in these examples on the computer, none of them were used in this book. The main intention of these solutions are to present all the basic steps for use with a calculator and identify the applicable equations. The approach presented for each example is not the only approach. Also, the statistical technique may not be the best or the easiest method to study the issue. These are extracted from proposals or studies that were successful or obtained favorable results. Care must be exercised in adapting these examples to your situations. The important thing to remember, as was discussed in Chapter 11, is the stages in the decision process, especially identifying the nature of your environment and variables. The cases are presented for the non-mathematical-minded EHS professionals who prefer simple scientific approaches to analyzing practical issues.

### 12.2 Example 1: Decision Tree and Expectation Values

Company ONE uses nuclear fission and fusion as part of its production process. The process and equipment are outdated. Within the last five years, the company was cited three times for safety violations. The average cost of the violations was $117,000. The company is considering improving the safety of its production by introducing a new patented process and pieces of equipment. The new pieces of equipment can be leased from the patent holder of the process for a maximum of four years. The annual cost of leasing the equipment and using the process depends on the production as shown:

| Unit A | < 600,000 items | $50,000 |
| Unit B | 600,000 to 800,000 | $100,000 |
| Unit C | >800,000 to 1,000,000 | $180,000 |

The introduction of these new pieces of equipment will result in the company not able to produce by-products that currently earn a profit $90,000 per annum. The production manager supported the project and recommended that the existing equipment be used in the final stage of the production process. The marketing manager presented a strong case against purchasing the new equipment. However, the safety manager used decision tree and expected values to support the hypothesis that the project is profitable. Based on marketing manager's report, the safety manager identified all the possible leasing options and supply as follows:

Unit A with annual supply of 600,000 items

Unit B with annual supply of 600,000 items

Unit B with annual supply of 800,000 items

Unit C with annual supply of 600,000 items

Unit C with annual supply of 800,000 items

Unit C with annual supply of 1,000,000 items

The probabilities of the supplies, projected contribution and expected contribution were tabulated as shown in Table 12.1 and presented a decision tree shown in Figure 12.1.

**TABLE 12.1**

Supply, Probability, and Expected Values of Example

| Supply | Probability | Projected Contribution per Item in $ | Expected Contribution $,000 |
|---|---|---|---|
| 600,000 | 0.80 | 2.05 | 984 |
| 800,000 | 0.65 | 2.50 | 1300 |
| 1,000,000 | 0.45 | 2.80 | 1260 |

The contributions from the current production are $1.82 for supply level less than 600,000 items, and $1.96 for supply level between 600,000 and 800,000 items. The average annual citation fines over the last five years is $70,000 with a probability of citation each year is 3 out of 5, or 0.6. Data from similar companies using the new process and equipment showed a probability of 0.01 of being cited. Using the decision tree, the safety manager advanced that these options are independent and only one can occur. Also, each option is equally likely to occur, hence equal probabilities. As a

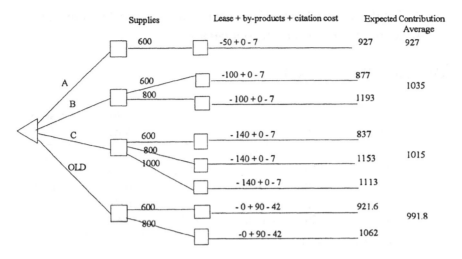

Figure 12.1   Decision Tree of Project Number One.

result, the expected contributions will be the arithmetic mean. The decision options are to lease either units A or B or C, or continue production.
The expected contributions are as follow:

| | |
|---|---|
| Lease unit A | $ 927,000 |
| Lease unit B | $1,035,000 |
| Lease unit C | $1,015,000 |
| Continue old | $991,800 |

The safety manager further developed a net present discounted cash
flow profile for the company to encourage management to purchase the
equipment instead of leasing. He established three financial options available to the company:

1. Full purchase of $50 M with a bank loan, repayable at the end
   of 4 years and with interest at 12% payable at the end of each
   year.
2. Continue leasing using lease option B and purchase for $100 M
   at the end of the 4th year.
3. Hire purchase with an initial payment of $15 M and 4 annual
   payments of $15 M.

Assuming the following as worse-case scenario:

1. The company uses leasing option B and the expected contributions remain constant over the four years.
2. The profit remains constant at $7.5 M annually.

3. The discount rate of 12% remains constant over the four years.
4. No major changes in the operations.

## Option to Purchase (a)

| Year Ending | Profit | Payment | Balance | Discount Value | Present Value |
|---|---|---|---|---|---|
| Start year 1 | | 0.00 | 0.00 | 1.00 | 0.00 |
| 1 | 7.50 | −6.00 | 1.50 | 0.89 | 1.34 |
| 2 | 7.50 | −6.00 | 1.50 | 0.80 | 1.20 |
| 3 | 7.50 | −6.00 | 1.50 | 0.71 | 1.06 |
| 4 | 7.50 | −56.00 | −48.50 | 0.64 | −31.04 |

Net present value = −$27.44 M.

## Option to Lease (b)

| Year | Profit | Payment | Balance | Discount Value | Present Value |
|---|---|---|---|---|---|
| Start year 1 | 0.00 | | | 1.00 | |
| 1 | 7.50 | −0.10 | 7.40 | 0.89 | 6.50 |
| 2 | 7.50 | −0.10 | 7.40 | 0.80 | 5.84 |
| 3 | 7.50 | −0.10 | 7.40 | 0.71 | 5.18 |
| 4 | 7.50 | −100.10 | −92.60 | 0.64 | −59.26 |

Net present value = −$41.74 M.

## Option to Hire Purchase (c)

| Year | Profit | Payment | Balance | Discount Value | Present Value |
|---|---|---|---|---|---|
| Start year 1 | 0.00 | 15.00 | −15.00 | 1.00 | −15.00 |
| 1 | 7.50 | 15.00 | −7.50 | 0.89 | −13.35 |
| 2 | 7.50 | 15.00 | −7.50 | 0.80 | −12.00 |
| 3 | 7.50 | 15.00 | −7.50 | 0.71 | −5.33 |
| 4 | 7.50 | 15.00 | −7.50 | 0.64 | −4.80 |

Net present value = −$50.48 M.

The option to purchase in year one was recommended. As a matter of interest the company purchased the equipment in year two.

## 12.3 Example 2: Sampling and Estimation

Failure of a major component used in the manufacturing of a game for children has resulted in major financial losses to the company. These financial losses were due to injuries to children and subsequent bad publicity. Company TWO manufactures many of two components in batches of 1000. These components must be produced to rigorous standards of accuracy. The finished component may be tested to discover whether they are defective at a cost of $1.25 per item. At this stage, if a component is defective, it can be corrected at a cost of $5.00. If the components are not tested before assembling, it is expected that any defects would become apparent later when the final product is tested. At this stage, if a component is determined to be defective, it will cost $10.00 to correct. The current practice is to randomly select 100 components per batch for testing. If more than 24% or 24 components are defective, the batch is rejected. The quality control department and production manager accused the safety engineering department of not doing a good job in building safety into the design. The safety manager used his knowledge of sampling and estimation to show that the sampling sizes of 100 per batch were insufficient to establish with 95% confidence level that less than 25% of the components are defective. The safety manager tested a random sample of 100 components from a recently completed batch. It was discovered that 20% of the components were defective. Using this information, the following calculations were done to decide:

1. The minimum proportion of defective components in a batch for it to be cheaper for the company to test all the components.
2. The sample size required to show with 95% confidence that the proportion of defective components in the population are less than 25%.

## Calculations

### Option 1: Testing all components

| | |
|---|---|
| The cost of testing a batch of components | $= 1000 * 1.25 = \$1250$ |
| Cost of correcting $p$ defective components in the batch | $= 5.00 * p * 1000 = \$5.00\, p$ |
| Total cost of testing and correcting | $= \$1250 + 5000\, p$ |

### Option 2: Not testing and correcting at assembling

| | |
|---|---|
| Total cost | $= 1000 * 10 * p = \$10,000\, p$ |

Testing all components would be cheaper, if option 1 is less than option 2, that is: $1250 + 5000\,p < 10{,}000\,p$. Therefore, $p > 0.25$, which is 25% components in a batch must be defective before it is cheaper to test all of them.

Reviewing the information collected from the sample tested and assuming normal distribution, the 95% confidence limits for a proportion are given by $p + 1.96\,s$. Where $s$ is the standard deviation and 1.96 is obtained from the normal distribution table. Portions of the normal curve are shown in Table 12.2.

**TABLE 12.2**

Proportions of Area under Normal Curve with $z$ Standard Deviations of Mean

| Proportions of Area | Standard Deviations ($z$) |
| --- | --- |
| 0.4 | 1.28 |
| 0.425 | 1.44 |
| 0.45 | 1.64 |
| 0.475 | 1.96 |
| 0.49 | 2.33 |

Number of standard deviations that will allow 95% of the normal distribution to be less than 0.25 is 1.64 to the right of the sample estimate of 20.

Using the formula $n = \{(1.64)^2\,pq\}/E^2$ where $E$ is the allowable error or the permissible risk,

$$n = \{(1.64^2)(0.2)(0.8)\}/0.05$$
$$= 172 \text{ components.}$$

Clearly a sample size of 172 is required to show with 95% confidence that the proportion of defective components are less than 25% assuming that the sample continues to show 20% defective. As result, the company quality-control department began randomly testing 175 components per batch. This was the solution of the problem since more batches were rejected and fewer games failed.

## 12.4 Example 3: Hypothesis Tests with Chi-Squared Distribution

The accident rates at THREE Company are always high. Five safety managers have been replaced within the last three years for incompetence. The company has been unionized and the union strongly opposes automation.

The plant manager has been assigned the additional responsibilites of the safety manager by the board of directors. Intuitively, he believes that the preference for any type of equipment depends on the accident experiences of the worker. Therefore, he consulted a representative sample of the work force to discuss the introduction of some new equipment. Four groups of workers were identified: group D with less than 5 accidents encounters, group C, <10 to 5, group B, <20 to 10, and group A >20. Three different types of equipment were discussed, resulting in three options:

1. Keep the existing semiautomatic equipment.
2. Introduce new full-automated equipment.
3. Introduce robotics in areas where the safety risk is high.

The options selected by the representative of the work force are shown in Table 12.3.

**TABLE 12.3**

Number of Employees in Favor of Option

| Equipment Favored | Group A | Group B | Group C | Group D |
|---|---|---|---|---|
| Semiautomatic | 18 | 10 | 40 | 27 |
| Fully automatic | 50 | 38 | 25 | 39 |
| Microchip | 22 | 32 | 35 | 44 |

He decided to test the hypothesis that there are differences among the four groups in the spread of opinion among employees at the 1% level of significance. The labor union proposed extensive training for the workers for acceptance of the proposal to automate. The union calculated the time needed to train employees in the use of the new automatic equipment is approximately normally distributed safely, with a mean of 7 days and a standard deviation of 2 days. This estimate was obtained from the following sample of 150 employees as shown in Table 12.4.

Management was unwilling to accept this condition. Management deemed that the sample is biased. To establish the plant manager's hypothesis the following calculations were done:

Let the null ($H_0$) hypothesis be: there is no difference among the 4 groups in the spread of opinion among employees.

Of the 380 employees sampled:
   95 (25%) preferred the semiautomatic equipment;
   152 (40%) preferred the fully automatic equipment;
   133 (35%) preferred microchip technology.

**TABLE 12.4**

Training Requirements of Employees

| Training Time (Days) | Frequency |
|---|---|
| 2 and < 3 | 3 |
| 3 and < 4 | 9 |
| 4 and < 5 | 12 |
| 5 and < 6 | 30 |
| 6 and < 7 | 45 |
| 7 and < 8 | 27 |
| 8 and < 9 | 12 |
| 9 and < 10 | 3 |
| 10 and < 11 | 6 |
| 11 and < 12 | 3 |

The null hypothesis can be translated to read that opinion in all groups is spread in these proportions. Constructing the contingency table (size $3 \times 4$) as follows: (O = observed, E = expected).

Contingency Table

| Preference | Group A O | Group A E | Group B O | Group B E | Group C O | Group C E | Group D O | Group D E | Total |
|---|---|---|---|---|---|---|---|---|---|
| Semiautomatic E = 25% | 18 | 22.5 | 10 | 20 | 40 | 25 | 27 | 27.5 | 95 |
| Fully automatic E = 40% | 50 | 36.0 | 38 | 32 | 25 | 40 | 39 | 44.0 | 152 |
| Robotics E = 35% | 22 | 31.5 | 32 | 28 | 35 | 35 | 44 | 38.5 | 133 |
| | 90 | 90.0 | 80 | 80 | 100 | 100 | 110 | 110.0 | 380 |

| O | E | O–E | $(O–E)^2$ | $(O–E)^2/E$ |
|---|---|---|---|---|
| 18 | 23 | –5 | 20 | 1 |
| 50 | 36 | 14 | 196 | 5 |
| 22 | 32 | –10 | 90 | 3 |
| 10 | 20 | –10 | 100 | 5 |
| 38 | 32 | 6 | 36 | 1 |
| 32 | 28 | 4 | 16 | 1 |
| 40 | 25 | 15 | 225 | 9 |
| 25 | 40 | –15 | 225 | 6 |
| 35 | 35 | 0 | 0 | 0 |
| 27 | 28 | –1 | 0 | 0 |
| 39 | 44 | –5 | 25 | 1 |
| 44 | 39 | 6 | 30 | 1 |

Sum of $(O–E)^2/E$ is equal to the chi-squared = 31.91.

There are $(m - 1)(n - 1) = (3 - 1)(4 - 1) = 6$ degrees of freedom. At the 1% significance level the critical point in the $\chi^2$ distribution for 6 degrees of freedom is 16.8. Since the value from the sample exceeds this amount, the null hypothesis can be rejected. The manager's view that there is a significance difference in the opinion of employees in different groups is plausible. As a result, he was successful in approaching the workers in groups based on their accidents encounters to accept automation.

To counteract the union's demand, the plant manager showed that the sample does not come from a population that is normally distributed with a mean of 7 days and a standard deviation of 2 days. He did this by first calculating the expected frequencies in a normal distribution with a mean of 7 and a standard deviation of 2, for the same class intervals as the union's sample. The probability of training time in less than 1 day was ignored since it was an insignificant amount. These are presented in Table 12.5

**TABLE 12.5**

Calculations of Probabilities Related to Example 3

| Class Interval Days | Days Below or Above Mean | Number of Standard Deviations Below or Above Mean | Probability of Less Than This Duration (from Tables) | Incremental Probability |
|---|---|---|---|---|
| <2 | –5 | –2.5 | 0.0062 | 0.0062 |
| 2 and <3 | –4 | –2.0 | 0.0228 | 0.0166 |
| 3 and <4 | –3 | –1.5 | 0.0668 | 0.0440 |
| 4 and <5 | –2 | –1.0 | 0.1587 | 0.0919 |
| 5 and <6 | –1 | –0.5 | 0.3085 | 0.1498 |
| 6 and <7 | 0 | 0.0 | 0.5000 | 0.1915 |
| 7 and <8 | +1 | +0.5 | 0.6915 | 0.1915 |
| 8 and <9 | +2 | +1.0 | 0.8413 | 0.1498 |
| 9 and <10 | +3 | +1.5 | 0.9332 | 0.0919 |
| 10 and <11 | +4 | +2.0 | 0.9972 | 0.0440 |
| 11 and <12 | +5 | +2.5 | 0.9938 | 0.0166 |
| over 12 | | 1.0000 | 0.0062 | |

The incremental probabilities in the right-hand column were multiplied by 150 ( the number of employees in the sample) to obtain the expected frequencies of training times if the sample were exactly normally distributed with a mean of 7 and a standard deviation of 2. These expected frequencies are shown in the Table 12.6

**TABLE 12.6**

Calculations of Example of $\chi^2$ Related to Example 3

| Class Interval Days | Observed Frequencies (O) | Expected Frequencies (E) | (O–E)² |
|---|---|---|---|
| <2 | 0 | 0.93 | 0.93 |
| 2 and <3 | 3 | 2.49 | 0.1 |
| 3 and <4 | 9 | 6.6 | 0.87 |
| 4 and <5 | 12 | 13.79 | 0.23 |
| 5 and <6 | 30 | 22.47 | 2.52 |
| 6 and <7 | 45 | 28.72 | 9.23 |
| 7 and <8 | 27 | 28.72 | 0.1 |
| 8 and <9 | 12 | 22.47 | 4.88 |
| 9 and <10 | 3 | 13.79 | 8.44 |
| 10 and <11 | 6 | 6.6 | 0.05 |
| 11 and <12 | 3 | 2.49 | 0.1 |
| over 12 | 0 | 0.93 | 0.93 |

$\chi^2$ is the sum of column four and equal to 28.38. There were 12 class intervals; therefore (12 – 3) = 9 degrees of freedom. At the 5% level of significance, for 9 degrees of freedom, the critical point in a $\chi^2$ distribution is equal to 16.9. Since the value from the sample exceeds this amount the null hypothesis was rejected, that is, there is no difference between the sample distribution and a normal distribution. The sample does not come from a population that is normally distributed with a mean of 7 days and a standard deviation of 2 days. As result the union's demand was unreasonable and denied.

## 12.5 Example 4: Hypothesis Tests with t-Distribution

Factory FOUR has 12 identical drilling machines. Failure to properly maintain these machines was a major cause of accidents. The machines all had safety guards installed, still accidents continued to occur. The safety officer inspected the lost-time records of the last ten years. The records showed that the total lost days followed a normal trend. The only detailed data available to the safety manager was the current year as shown in Table 12.7. As part of the safety manager's preparation of the annual budget, she used the existing data to estimate the expected number of lost days per machine in the coming year. Using this information, she requested an additional sum of 50% of the cost of lost days to carry out a safe-behavior program. As a matter of interest, management approved it and lost days in the subsequent year declined by 63%.

**TABLE 12.7**

Lost Time per Machine, 1993

| Machine No. | Days Lost |
|:---:|:---:|
| 1 | 16 |
| 2 | 12 |
| 3 | 21 |
| 4 | 17 |
| 5 | 15 |
| 6 | 12 |
| 7 | 18 |
| 8 | 18 |
| 9 | 14 |
| 10 | 20 |
| 11 | 16 |
| 12 | 19 |

The safety manager calculated the estimated average lost days as follows:

Mean lost days = 198/12 = 16.5 days

| | Variance | |
|:---:|:---:|:---:|
| $x$ | $x - \bar{x}$ | $(x - \bar{x})^2$ |
| 12 | −4.5 | 20.25 |
| 21 | 4.5 | 20.25 |
| 17 | 0.5 | 0.25 |
| 15 | −1.5 | 2.25 |
| 12 | −4.5 | 20.25 |
| 18 | 1.5 | 2.25 |
| 18 | 1.5 | 2.25 |
| 14 | −2.5 | 6.25 |
| 20 | 3.5 | 12.25 |
| 16 | −0.5 | 0.25 |
| 19 | 2.5 | 6.25 |
| Total | 198 | 93.00 |

Mean of the sample = 198/12 = 16.5 lost days

The variance of the sample is 93/12 = 7.75 lost days

The best estimate of the population variance = 7.75 * 12/(12 − 1) = 8.4545 days, therefore the standard deviation = $\sqrt{8.4545}$ = 2.91 days

Therefore, the standard error of the sample mean $(\sigma / \sqrt{n}) = 2.91 / \sqrt{12}$ = 0.84 lost days

For 95% confidence level for a t distribution where there is (12 − 1) = 11 degrees of freedom and at the two-tail test, t = 2.20. Therefore, the projected estimated mean number of days loss due to accidents per machine at 95% confident level = 16.5 ± 2.20 (0.84) days.

= 16.5 ± 1.85 days or, 14.65 to 18.35 days.

## 12.6 Example 5: Linear Programming

Company FIVE produces two products: M and R. The tendency of the company is to produce as many as possible Rs because of the bigger profit margin. Within the last twenty months, the average production of R was 325 units. Most of the accidents occurred in the production of R. Despite many requests from the safety manager to decrease the production of R, management refused to accommodate the request for fear of loss profit. The safety manager collected data and did a linear programming exercise to show that the same profit or more can result from producing less of R. The cost data relating to the production of the two products M and R are as follows:

| Cost Factors | M | R |
|---|---|---|
| Selling price | $150 | $100 |
| Direct cost | $80 | $30 |
| Indirect cost | $30 | $20 |

Production information: each unit of products incurs costs of machining and assembly. The total capacity budgeted were 700 hours of machining and 1000 hours of assembly per month. The cost of this capacity was fixed at $7000 and $10,000 respectively per month, whatever usage made of it. The number of hours required in each of these areas to complete one unit of output is as follows:

| | M | R |
|---|---|---|
| Machining | 1.0 | 2.0 |
| Assembly | 2.5 | 2.0 |

Under the terms of special controls introduced by the government, the selling prices are fixed: the maximum output is 400 units and minimum is 100 for either product. At the present selling prices, demand exceeds the supplies. When safety is considered, the number of accidents per products per manufacturing stage is as follows:

| | Number of Accidents | |
|---|---|---|
| Product | Machining | Assembly |
| M | 45 | 36 |
| R | 120 | 23 |

The average cost of an accident in the machining area is $23.50 and $13.75 in the assembly area. The optimal production was first calculated as follows:

Let the number of units of M produced be $x$

Let the number of units of R produced be $y$

The optimal production plan is assumed to be one that maximizes contribution and profit, hence the objective function is: maximize $40x + 50y$

Constraints:

| | | |
|---|---|---|
| $x + 2y$ | $\leq 700$ | (machining hours) |
| $2.5x + 2y$ | $\leq 1000$ | (assembly hours) |
| $x$ | $\leq 400$ | (maximum output) |
| $y$ | $\leq 400$ | (maximum output) |
| $x,y$ | $\geq 100$ | (government agreement) |

Plotting the information on a graph as shown in Figure 12.2, the maximum contribution occurs at point P on the graph, where:

$2.5x + 2y = 1000$ and $x + 2y = 700$
Solving $x = 200$ and $y = 250$.

Comparison of plans:

| Product | $ Contribution per Unit | Current Plan Units | Contribution | New Plan Units | Contribution |
|---|---|---|---|---|---|
| M | 40 | 100 | 4000 | 200 | 8000 |
| R | 50 | 325 | 16,250 | 250 | 12,500 |
| | Total | | $20,250 | | $20,500 |

Also, the total number of machining hours will decrease from 750 to 700. This will evidently result in fewer accidents, so larger profit margins

---

## 12.7 Example 6: Network Analysis

Company SIX is a construction company. Ten years ago there was a fatality at a construction site. The safety director was held responsible and charged for a civil offence. Completing his prison term, he went back to school and pursed a degree in mathematics. Using his knowledge on network analysis, he examined the project that led him to be sentenced to prison. He collected information on the company's last twenty projects. He used this information to break the fatal project into the following eight independent activities and probabilities of completion shown in Table 12.8.

**TABLE 12.8**

Project Activities and Probabilities of Completion

| | | Estimated Duration | | |
| | | | Most | |
| Activity | Preceding Activity | Optimistic | Likely | Pessimistic |
|---|---|---|---|---|
| A | — | 4 | 11 | 12 |
| B | — | 45 | 48 | 63 |
| C | B | 13 | 33 | 35 |
| D | B | 25 | 29 | 39 |
| E | A, C | 14 | 21 | 22 |
| F | D, E | 18 | 32 | 34 |
| | A, C | 17 | 19 | 27 |
| H | G | 15 | 20 | 25 |

He noted that the project was scheduled to be completed in a maximum of 120 days. The fatal accident occurred on the 99th day. The company policy did not permit the hiring of additional workers. Most workers were not interested in overtime because they worked 6 days a week at an average of 54 hours and company policy on overtime that only paid overtime pay only after working 60 hours per week. The ex-safety director developed a network as shown in Figure 12.2. He used PERT to analyze his network as follows.

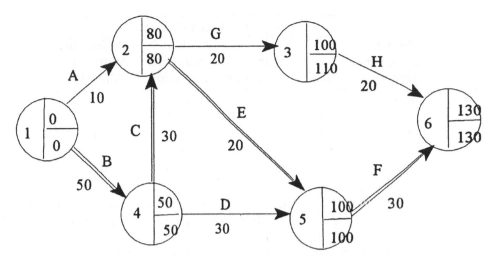

**Figure 12.2   Network of Case Number Six.**

The mean, standard deviation, and variance times are as shown in Table 12.9. These were calculated by applying Equations 10.1 and 10.2 and substituting the optimistic, most likely, and pessimistic times given.

TABLE 12.9

Mean and Standard Deviation of Times for Example 6

| Activity | Mean Time | Standard Deviation |
|----------|-----------|--------------------|
| A | 10.00 | 1.33 |
| B | 50.00 | 3.00 |
| C | 30.00 | 3.67 |
| D | 30.00 | 2.33 |
| E | 20.00 | 1.33 |
| F | 30.00 | 2.67 |
| G | 20.00 | 1.67 |
| H | 20.00 | 1.67 |

From the network it was seen that the critical path is B, C, E, F giving an expected project time of 130 days. If the distribution of the total project time is normal, the probability of the project time being greater than 120 days was 97.75%. This was calculated as follows:

| Routes Through the Network | On Route (Days) | | No. Standard Deviations | Probability Within 120 Days (%) |
|---|---|---|---|---|
| | Mean | Standard Deviation | | |
| BCEF | 130 | 5.6 | −1.78 | 3.75 |
| BDF | 110 | 4.6 | 2.17 | 98.5 |
| BCGH | 120 | 5.3 | 0 | 100 |
| AEF | 60 | 3.3 | 18.18 | 100 |
| AGH | 50 | 2.7 | 25.92 | 100 |

The probability that the project would exceed 120 days is the joint probability of routes BCEF and BDF, which is equal to 96.25 + 1.5 = 97.75%. As result of this analysis, the ex-safety director established that completing this project in the scheduled time without workers taking risk was impossible. He sued the company and won over $2.1 M in compensation for his wrongful detention.

# Selected Bibliography

Ott, Wayne R., *Environmental Statistics and Data Analysis*, Lewis Publishers, Boca Raton, FL, 1995.

Zwillinger, Daniel, ed., *Standard Mathematical Tables and Formulae*, 30th ed., CRC Press, Boca Raton, FL, 1996.

Petersen, Dan, *Techniques of Safety Management: a Systems Approach*, 3rd Ed., Aloray, New York, 1989.

Selvin, Steve, *Statistical Analysis of Epidemiologic Data*, Oxford University Press, New York, 1991.

Petitti, Diana B., *Meta-Analysis, Decision Analysis, and Cost-Effectiveness Analysis*, Oxford University Press, New York, 1994.

Hildebrand, Stephen G. and Johnnie B. Cannon, *Environmental Analysis: the NEPA Experience*, Lewis Publishers, Boca Raton, FL, 1993.

Joseph, Anthony, *Health, Safety, and Environmental Statistics*, U.W.I., Trinidad, 1991.

Anderson, David R., Dennis J. Sweeney, and Thomas A. Williams, *Statistics for Business and Economics*, West Publishing, Minneapolis/St. Paul, 1993.

Cothern, C. Richard and N. Phillip Ross, eds., *Environmental Statistics, Assessment, and Forecasting*, Lewis Publishers, Boca Raton, FL, 1994.

Kavianian, H. R. and C. A. Wentz, Jr., *Occupational and Environmental Safety Engineering and Management*, Van Nostrand Reinhold, New York, 1990.

IT Environmental Programs and ICF Kaiser International, *Guidelines for Statistical Analysis of Occupational Exposure Data*, Office of Pollution Prevention and Toxics, U.S. Environmental Protection Agency, Washington D.C., 1994.

Chartered Institute of Management Accountants, *CIMA Study Text: Stage 1, Quantitative Methods*, 4th Ed., BPP, London, 1990.

Chartered Association of Certified Accountants, *ACCA Study Text: 2.6 Quantitative Analysis*, 1st Ed., BPP, 1988.

# *Index*